Skylab

A GUIDEBOOK

by
LELAND F. BELEW
and
ERNST STUHLINGER
GEORGE C. MARSHALL SPACE FLIGHT CENTER
NATIONAL AERONAUTICS AND SPACE ADMINISTRATION

Skylab

A GUIDEBOOK

Skylab

A GUIDEBOOK

Acknowledgments

This document describes the result of the work of several thousand engineers and scientists who, over the last ten years, have conceived, designed, developed, built, and tested Skylab, the most complicated space system in the American space flight program so far. As we are completing our writing just a few months before the launching of Skylab, we extend our appreciation and gratitude to all of those who have supported us in writing this booklet. Members of many organizations at NASA Headquarters in Washington, at the Lyndon B. Johnson Space Center, the Goddard Space Flight Center, the John F. Kennedy Space Center, and the George C. Marshall Space Flight Center have provided valuable advice and help. The Life Sciences Directorate at the Lyndon B. Johnson Space Center and the Office of Life Sciences at NASA Headquarters made substantial contributions to the chapter on Life Sciences Projects. Dr. G .C. Bucher, M. I. Kent, and R. M. Nicholson of the G. C. Marshall Space Flight Center were rsponsible for organizing the material and composing most of the booklet in its present form.

LELAND F. BELEW
Manager, Skylab Program Office
MSFC, Huntsville, AL

ERNST STUHLINGER
Associate Director for Science
MSFC, Huntsville, AL

Preface

Skylab will be the first manned project in the space program of the United States with the specific purpose of developing the utility of space flight in order to expand and enhance man's wellbeing on Earth. Experiments and observations will be carried out on Skylab in a number of different areas, among them observations of the Earth, solar astronomy, stellar astronomy, space physics, geophysics, biomedical studies, zero-gravity biological studies, zero-gravity technology studies, and spacecraft environment.

In all these areas, exciting observations and discoveries already have been made during the brief history of space flight. However, Skylab will offer decisive improvements over earlier space projects. It is an experimental space station that will provide a comfortable shirtsleeve environment for its three-man astronaut-scientist crew. Three crews will occupy Skylab in three shifts for periods of 28, 56, and 56 days. During Skylab's eight-month operational lifetime, some instruments will work in automated modes, transmitting their data directly to Earth or storing them on tapes or film. Other instruments will be operated or monitored by the astronauts. In these observations, man will contribute his unique capabilities for on-the-spot judgment, decision-making, analyzing and interpreting unusual situations, recognizing unexpected developments, changing the course of an experiment, selecting a new target of study, adjusting and reorienting instruments, learning quickly from experience, adapting readily to new conditions, and discussing his observations with specialists on the ground. These qualities, so valuable to the scientist on Earth, will give the scientist in space the power to advance research far beyond the limits set by the earthly environment. The Skylab mission will utilize man as an engineer and as a research scientist, and it will give him the opportunity of assessing his potential capabilities for future space missions.

Skylab will have several distinct goals: to enrich our scientific knowledge of the Earth, the Sun, the stars, and cosmic space; to study the effects of weightlessness on living organisms, including man; to develop methods for the processing and manufacturing of materials utilizing the absence of gravity; and, in perhaps its most important objective, to develop means of observing and monitoring the Earth's surface in support of earthly needs. In addition, a program of student projects will be carried out on Skylab, designed to stimulate interest in science and technology among high school students. More than 3,400 boys and girls in secondary schools submitted proposals for space experiments and demonstrations on Skylab; 19 of these proposed experiments were selected to be carried out during the Skylab missions.

With a total length of about 35 meters (117 ft.),[1] Skylab has the size of an average house. Its launch weight is 90,606 kilograms (199,750 lbs.).[2] Skylab's orbital altitude will be 432 km (268 statute miles or 234 nautical miles);[3] it will orbit the Earth once every 93 minutes. With its relatively steep orbital inclination of 50 degrees, it will be visible from the earth under lighting conditions which prevail during several hours after sundown and before sunrise, in all regions other than the Arctic and Antarctic. The orbital trajectories of Skylab will sweep out an area which covers 75 percent of the Earth's surface, 80 percent of its food-producing regions, and 90 percent of its population.

Skylab will represent a milestone of paramount importance in the American space program. It may turn out to be the beginning of man's permanent foothold and settlement in space. This booklet will give a brief overview of Skylab, its history, its design, its operation, and its program of experiments and observations.

Skylab is the most diversified and most complex orbiting spacecraft of the U.S. space program so far; its potential returns in experience and knowledge will be of decisive value to science, and they will help us in our effort to cope with many of the problems which beset man's life on Earth.

---

[1] 1 meter=1 m=3.28 ft=39.37 in.
[2] 1 kilogram=1 kg=2.2 lb.
[3] 1 kilometer=1 km=0.622 statute mile=0.54 nautical mile.

Contents

CHAPTER I

Science From Orbit—
An Introduction to Skylab

1. AN OUTPOST IN SPACE

"Unlimited vacuum of outer space and the absence of gravitational forces in an orbiting satellite," said Professor Hermann Oberth [1] in 1923, "make an Earth-circling spacecraft an ideal site for the observation of stars, the Moon, the Sun, the planets, and particularly our Earth." Barely 35 years later, these prophetic words began to transform into reality. Today, hundreds of instruments already have observed the sky and the Earth from satellite orbits. More than 80 astronauts and cosmonauts have seen our earth from vantage points above the atmosphere and from as far away as the moon. Impressive discoveries were made with unmanned probes and satellites; among them X-rays from the sun and numerous other stellar objects; the radiation belts around the Earth, the magnetosphere, [2] the solar wind, and the geocorona, not to speak of the vast improvement of our capabilities for meteorological and other Earth-oriented observations. Apollo proved the usefulness and efficiency of man in space as a pilot and navigator, a technician, a troubleshooter, and particularly as an observer and a research scientist.

Our first projects in space provided us with a harvest of scientific understanding and of technological experience beyond any expectations. The value of the space program for the enrichment of life on Earth has become manifest in three distinct ways: space flight presents a new frontier of exploration; it opens the doors to a huge source of scientific knowledge; and it offers direct help in many of our earthly needs. Early pioneers of space flight, anticipating these benefits from Earth-orbiting outposts, have often suggested that long-time stations in orbit be established. These stations would be equipped with instruments to observe the sky and the Earth, with receivers and transmitters for global communications, and with processing machinery which utilizes the state of weightlessness; they would be manned by astronauts who use, operate, and service the instruments with an efficiency and a flexibility not attainable with automated equipment. An orbiting station would operate for years, while its crew members would be rotated every few weeks or months. As space flight began to evolve about 25 years ago, interest and work concentrated at first on high altitude rocket probes,

---

[1] Professor Hermann Oberth, Hungarian-born rocket expert and pioneer of space flight, lives in Feucht, West Germany.

[2] Scientific and technical terms are explained in the Glossary, page 237.

1

and then on small automated satellites. Plans for manned orbiting stations had to wait. However, with Apollo, large Earth-to-orbit transportation systems and life support facilities were developed. Instruments for the observation of astronomical objects, of the space environment, and of the Earth became available from numerous science and applications projects. Solar-electric power supplies, communication equipment, data systems, and attitude control systems were developed and operated from the early 1960's on. Conditions for the development of a manned orbiting space station, therefore, become favorable during the past decade, and plans for a first station in orbit began to materialize early in the sixties. Project work was started a few years later. Named "Skylab" in 1970, the first manned space station of the U.S. will be ready for launch in May, 1973. In its large and comfortable interior, three-man crews will conduct an exciting program of experiments and observations. Several of the experiments were developed by scientists in foreign countries. NASA has invited world-wide participation in the analysis and interpretation of Skylab data.

2. OPERATION ABOVE THE EARTH'S ATMOSPHERE

The shell of atmospheric gases which surrounds the earth as a source of oxygen and carbon dioxide for animals and plants, and as a protection against the dangerous radiations and the temperature extremes of space, sets a narrow limit to Earth-bound observations of celestial objects because it allows only a small portion of the total wavelength spectrum to reach instruments near the Earth's surface. The light-scattering effect of the atmosphere generates a background in the viewing field of an Earth-bound telescope which sets a lower limit of about the 24th magnitude to the star images which can be recorded from Earth. Furthermore, even the finest telescopes on the ground can provide only limited optical resolution because of the image-blurring effect of the atmosphere. Telescopes outside the Earth's atmosphere are not subject to these limitations. Several solar telescopes on Skylab will view the sun with a clarity and resolution, and in ultraviolet and X-ray regions of the spectrum, which would be impossible from the Earth's surface. These observations will substantially increase our knowledge of the Sun, of its mechanisms to convert and to transport energy, of the violent outbursts of radiations and particles in solar flares, of the strange behavior of hot plasmas, and of the many ways in which the Sun influences our weather, our environment, and, in fact, our lives on Earth. Skylab's stellar cameras will greatly expand the capabilities of Earth-based observatories by extending the range of measurements far into the ultraviolet regions which are inaccessible from Earth. Cosmic ray particles which would not be able to penetrate the Earth's atmosphere will be recorded during Skylab's flight with emulsion plates and plastic detectors; meteoritic particles will be collected on ultra-clean surfaces. Special optical sensors will observe the airglow in the highest layers of the atmosphere, and also the "Gegenschein" [1] which is probably caused by a reflection of light from dust clouds orbiting the Sun.

---

[1] Faint glow in the night sky in a direction opposite to the Sun.

3. ZERO GRAVITY ENVIRONMENT

On an orbiting spacecraft, the force of gravity, omnipresent in our earthly environment, is counteracted by centrifugal force. The resulting force is exactly zero at the center of gravity of the spacecraft. At other points within the spacecraft, the small resulting force varies between plus and minus about one millionth of the Earth's gravitational force at the surface, depending upon the distance and direction from the center of gravity.

The state of weightlessness on Skylab will be utilized by a number of experiments which study the influence of gravitational forces in biological, chemical, physical, and metallurgical processes. The cleavage of living cells may be triggered by a density stratification caused by gravity; cell metabolism may depend on gravitational forces; the content of minerals in the bone structure of animals and man may be influenced by gravity; vestibular functions, pressure distribution within the body, muscle tension, blood circulation, perhaps even the function of endocrine glands should be expected to depend on the force of gravity. The study of these functions under weightlessness will help our understanding of the mechanisms that govern many of the processes in living organisms.

Gravitational forces cause stratification of liquids which contain components of different densities. The desired mixing of certain alloys in the molten state, or of chemical reagents, is often difficult or even impossible on the Earth's surface because of this gravitational separation effect. Under conditions of zero gravity, fluids can be maintained in a state of perfect mixing. Even the mixing of liquid metals with gases, resulting in foams, can be achieved; the production of foam metals with extreme strength-to-weight ratios may be one of the future activities on orbiting space stations. The growth of single crystals from solutions, likewise, is often perturbed by effects attributable to gravity. These perturbing effects will not exist in orbiting laboratories, and single crystals of high purity may be obtainable in much larger sizes under weightlessness than on Earth. Such crystals are of great importance in the manufacturing of semiconductor components for electronic circuits.

4. BROAD VIEW OF THE EARTH'S SURFACE

Skylab will fly over about 75 percent of the Earth's surface. Each of its 93-minute orbital tracks will be repeated every five days. Photographic, infrared, and microwave equipment will provide pictures and measurements of the terrain underneath the spacecraft. The decisive advantage of viewing from a satellite orbit, rather than from airplane altitude, lies in the fact that a satellite sees a large area on Earth at the same time and under the same lighting conditions. Comparison of different parts of an area with regard to cloud cover, plant growth, irrigation, land use, urban expansions, crop conditions, natural resources, air pollution, water management, topographical features, geological formations, and other surface properties will be much easier with the "synoptic" [1] capability of a satellite. As an example, Figure 1

---

[1] "Synoptic" is the capability of seeing several different things at the same time and under the same conditions.

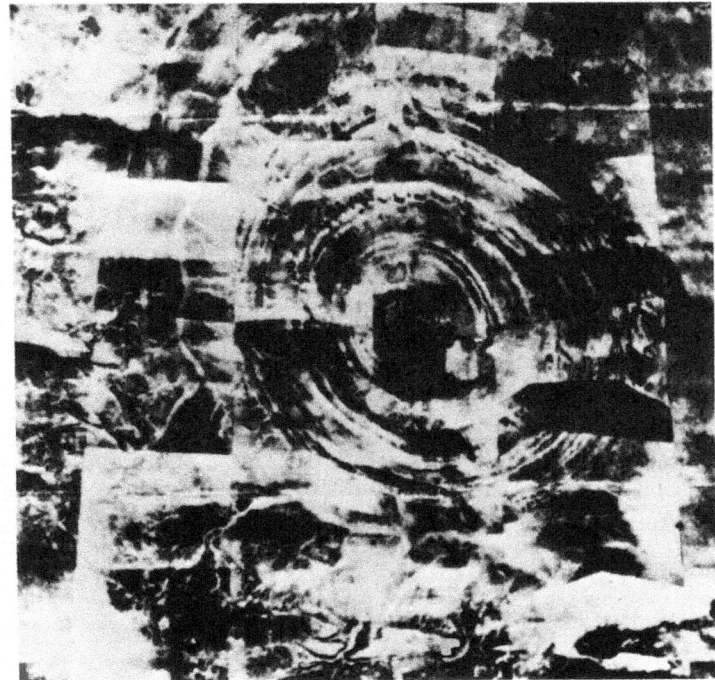

FIGURE 1.—Huge volcanic structure in Mauritania in Central Africa. (*left*) Mosaic from a number of airplane photographs. (*right*) Photograph from a Gemini mission. Comparison of the two pictures shows the superiority of satellite viewing over airplane viewing.

4

shows two pictures of the same surface feature, a huge volcanic structure in Mauritania in Central Africa. The first picture (a) is a composite of numerous photographs taken from an airplane; the second (b) was taken from a satellite. The superiority of satellite pictures is obvious. This superiority is further enhanced by the rapidity with which satellite pictures can be taken and transmitted to Earth.

One of the picture-taking systems on the Earth Resources Technology Satellite, launched in July, 1972, is capable of transmitting a photograph of the Earth every 25 seconds, covering a square of 185 by 185 km (115 by 115 statute miles), with a resolution of about 30 meters (100 ft). One of the Skylab cameras will cover squares of 109 km by 109 km (68 by 68 statute miles) with a resolution of 11.5 meters (35 ft). Skylab, during its three mission periods, will take almost 40,000 photographs with its several cameras. In conjunction with aircraft and ground-based observations, these photographs will allow an analysis and assessment of man's interaction with his environment unprecedented in previous studies.

CHAPTER II

History of the Skylab Program

From the beginning of the Apollo program, possibilities were considered to use Saturn Apollo components for future missions and applications after the lunar landings (Table 1).

During the early sixties, the Apollo Extensions Support Study identified possible new or modified flight projects which could use launch vehicles

TABLE 1.—*History of Skylab Program.*

1960	
1961	STUDIES FOR APPLICATION OF APOLLO COMPONENTS TO OTHER PROJECTS
1962	
	CONCEPT OF APOLLO TELESCOPE MOUNT (ATM)
1963	
	BEGINNING OF INSTRUMENT DEVELOPMENT FOR ATM
1964	
1965	APOLLO APPLICATIONS OFFICE ESTABLISHED
1966	"EXPERIMENT MODULE" WITH ATM
1967	"CLUSTER" CONCEPT
1968	"WET" WORKSHOP
1969	APOLLO 11 LANDING ON MOON "DRY" WORKSHOP
1970	NAME "SKYLAB"
1971	DEVELOPMENT, FABRICATION, AND TESTING OF SKYLAB
1972	
1973	SKYLAB LAUNCH

and spacecraft components developed for Apollo. One of the possibilities considered at that time was the use of an Apollo Command and Service Module (CSM) to carry an assembly of small solar telescopes into orbit, to deploy and operate them on the Service Module with the assistance of the astronauts, and to return the exposed films to Earth on board the Command Module (Fig. 3). This assembly was named Apollo Telescope Mount (ATM) in 1963. From these early efforts to extend the use of Apollo for further applications, a permanent organization evolved. On August 6, 1965, the Apollo Applications Office was established at NASA Headquarters in Washington, D.C. The Apollo Applications program was to include long duration Earth orbital missions during which astronauts would carry out scientific, technological, and engineering experiments. After completion of the Apollo program, spacecraft and Saturn launch vehicles originally developed for the lunar exploration program would be modified to provide the capability for crews to remain in orbit for extended periods of time.

As these studies progressed, plans for more elaborate observations of the Sun with a group of solar telescopes mounted on Apollo-related spacecraft also developed further. A first schedule, established in March, 1966, envisioned three Experimental Modules consisting of Saturn S–IVB spent stages which would be converted to "Workshops," and four Apollo Telescope Mounts. The Saturn S–IVB stage serves as second stage of the Saturn IB launch vehicle and also a third stage of the Saturn V launch vehicle.

According to these plans, a workshop in orbit would be installed in the empty S–IVB stage. This stage would ascend into space as part of the Saturn IB launch vehicle, carrying a manned Command and Service Module. After the S–IVB stage had used up its fuel, the astronauts in the CSM would dock with the stage and enter the stage's hydrogen tank through an airlock passageway. A number of biomedical experiments would be performed in the Command Module on this mission. No crew quarters were planned at that time in the S–IVB stage workshop; activities would

FIGURE 3.—Early 1960's concept of a telescope mount on the Command and Service Module.

be limited to familiarization with moving about in a controlled and enclosed environment under zero gravity. This concept of using the S–IVB stage was the precursor of the present Skylab.

On July 7, 1966, NASA announced the establishment of new Apollo Applications Program Offices at the Lyndon B. Johnson Space Center and the George C. Marshall Space Flight Center. A new schedule, released on December 5, 1966, called for launches of two Saturn IB vehicles about one day apart; the first unmanned, the second manned. The astronauts would make the S–IVB stage of the first vehicle habitable by installing equipment and introducing a life-supporting atmosphere in the stage's hydrogen tank so they could live and work there without the need for space suits. The hydrogen tank would be equipped before launching with two floors, some basic equipment, and an inner wall.

An Airlock Module (AM) would be attached to the S–IVB stage, and a Multiple Docking Adapter (MDA) would provide the attaching point for the Command Module carrying the astronauts to the stage. The stage, the Airlock Module, and the Multiple Docking Adapter constituted this first concept of the Orbital Workshop.

This plan also introduced the "cluster concept" which envisioned additional components to be attached to the Workshop. A modified Lunar Module ascent stage (LM) of the kind that carried astronauts from the Moon's surface in the Apollo Program and an Apollo Telescope Mount (ATM) would be launched together on one vehicle. The LM would be the control center for the ATM in orbit. This first launch would be followed by a manned launch. Lunar ascent stage and ATM would be attached to the Workshop at a docking port on the side of the Multiple Docking Adapter. The Command and Service Module with the astronauts would dock at the end of the docking port of the MDA.

This Workshop was called the "Wet Workshop" because the S–IVB stage to be used for the Workshop would be launched "wet," that is, filled with fuel to be consumed before reaching orbit. The empty hydrogen tank would be purged of remaining fuel and then filled with a life-supporting atmosphere.

In March, 1967, it was decided that the Orbital Workshop would have solar panels to produce electric power. This increase in electric power production was required as a result of astronauts living in the Workshop. Before this change was made, the Command and Service Module was planned to provide the Workshop's power except for the Apollo Telescope Mount which was to have its own solar-electric power supply.

Limited funds for the Apollo Applications Program led to a reduction of the number of launches and the extension of launch dates further into the future.

A major redirection of the Apollo Applications Program effort was made on July 22, 1969, six days after the first lunar landing. NASA announced plans to launch the Workshop and an elaborate Apollo Telescope Mount together on a two-stage Saturn V. The S–IVB (third) stage would not carry any fuel (hence the term "Dry Workshop") but would be completely equipped on the ground as a habitable system to be entered by the astronauts in orbit. This decision was aided by the successful lunar landing which made Saturn V launch vehicles available for other purposes.

Plans for two Saturn V launches with two Workshops and two ATM's, and for seven Saturn IB launches, were announced in 1969. The first Workshop launch was planned for July 1972.

The program was renamed on February 24, 1970, when the Apollo Applications Program became the Skylab Program. The Skylab cluster was to consist of the S–IVB Orbital Workshop, Airlock, Multiple Docking Adapter, and Apollo Telescope Mount. Early in 1971, the planning date for the launch was set for April 30, 1973.

The Skylab Program Office in the Office of Manned Space Flight in NASA Headquarters is responsible for overall management of the program (Fig. 4). The Marshall Space Flight Center (MSFC) at Huntsville, Alabama, has responsibility for developing and integrating most of the major components of the Skylab including the Orbital Workshop, Airlock Module, Multiple Docking Adapter, Apollo Telescope Mount, Payload Shroud, and most of the experiments. Further, MSFC has overall systems engineering and integration responsibility to assure the compatibility and integration of all system components for each flight. During and after the Skylab mission operations, MSFC will provide support for launch and flight operations. MSFC is also responsible for the Saturn IB and Saturn V launch vehicles.

The Lyndon B. Johnson Space Center (JSC) at Houston, Texas, is responsible for flight operations, recovery, crew selection and training, assigned experiments, and development of the modified Command and Service Modules and the Spacecraft Launch Adapter (SLA). In addition,

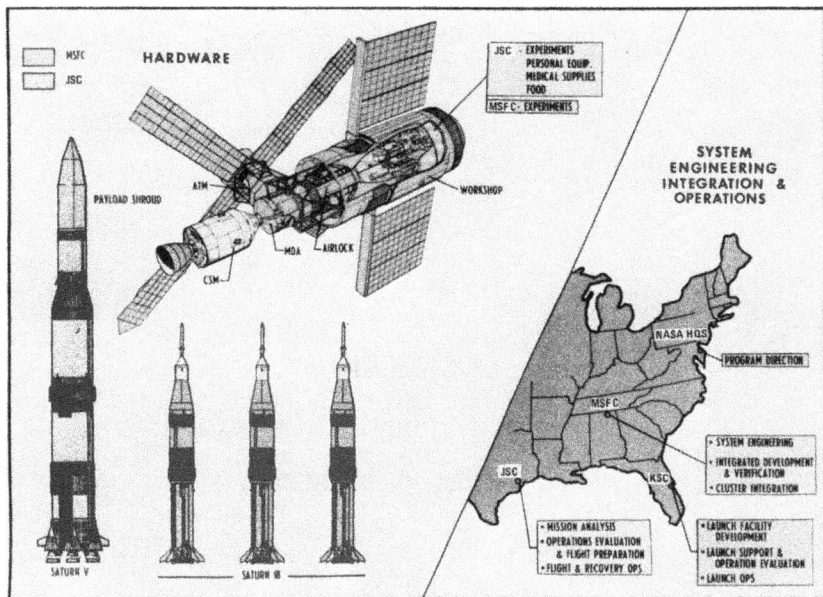

FIGURE 4.—Management responsibilities for Skylab Program.

9

JSC developed crew systems, medical equipment, food, and other crew-supporting components, and provided for stowage in the CSM of experimental data and samples to be returned from orbit. JSC also is responsible for overall mission analysis and evaluation, particularly from the viewpoint of crew performance.

The John F. Kennedy Space Center (KSC) in Florida is responsible for launch facilities for all Skylab flights, checkout procedures, and the planning and execution of launch operations, two of them within 24 hours.

Most of the design and manufacturing for Skylab was done by a large number of industrial companies under contract either directly to NASA, or as subcontractors to one of the major contractors. Fig. 5 shows the major Skylab contracts managed by NASA Centers. However, an extremely close cooperation between NASA, scientific investigators, and industry, extending over all levels of the program structure, existed from the beginning of the project. In fact, the work teams were so tightly integrated that it often would have been difficult to differentiate within these teams between contractors, NASA members, and university-type scientists as the project proceeded from its early concepts through development, manufacturing, testing, final assembly, and checkout.

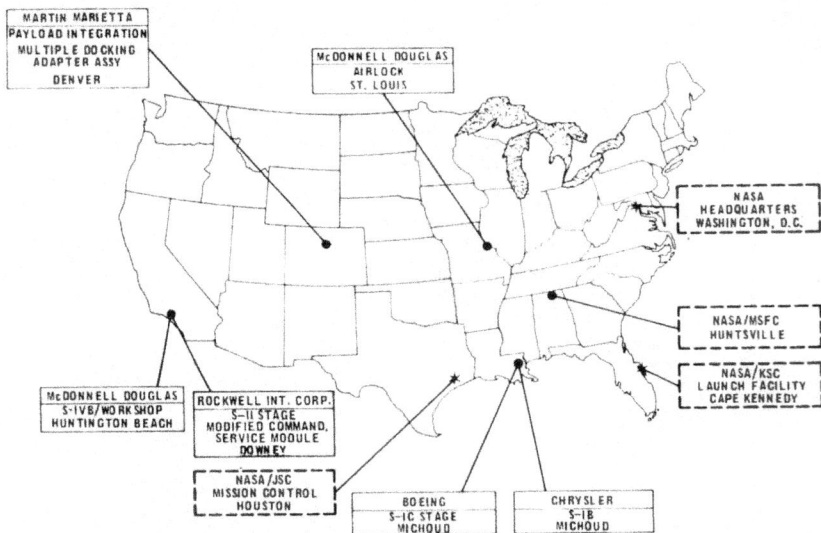

FIGURE 5.—Major Skylab contractors.

CHAPTER III

Profile of the Skylab Mission

Planning and early design work for Skylab started at a time when Project Mercury had ended and when the Gemini missions were beginning to accumulate experience in manned space flight. As plans for the project evolved during the 1960's, a number of other space projects of that period provided flight experience, technical data, and scientific results that proved most valuable for Skylab. In particular, the flights of the Apollo Program helped shape the profile of the Skylab mission. More than any other space flight project so far, Skylab encompasses a variety of mission elements of considerable complexity, among them the longest periods of weightlessness for the astronauts, the manned operation of a sophisticated solar observatory, a series of engineering tasks, scientific experiments inside Skylab, observations of the Earth, and biomedical studies by and of the astronauts. More than 270 different scientific and technical investigations will be supported by Skylab during its eight-month lifetime. In this chapter, the broad objectives of Skylab, the principal features of the huge spacecraft, and the basic plan for mission operations will be described.

1. MISSION OBJECTIVES

Skylab has the prime purpose of making spaceflight useful for man's endeavors on Earth. It will put knowledge, experience, and technical systems developed during the Apollo Program in service for a wide range of scientific and technological disciplines. Elements of spaceflight systems which have proven their capabilities on Apollo flights include propulsion systems, space power sources, guidance and control systems, communications and data systems, scientific sensors, life support system, Earth return capability, and ground support equipment. Skylab, representing the next big step in spaceflight development, has integrated these proven elements into a space system whose purpose is the practical utilization of space flight for earthly needs. These earthly needs cover a broad spectrum of human activities, including the expansion of our scientific knowledge in physics and astronomy, the study of our celestial environment, the production of new materials, the observation and monitoring of the Earth's surface, and research on living organisms, including man, under weightlessness.

Design and operation of Skylab aims at the following objectives (Table 2):

TABLE 2.—*Major Objectives of Skylab.*

Conduct Earth resources observations.

Advance scientific knowledge of the Sun and stars.

Study the processing of materials under weightlessness.

Better understand manned space flight capabilities and basic biomedical processes.

• Conduct Earth resources observations.

Growth of the population and improvement of its average living standard require a continuous increase of the efficiency and also the care with which the resources of the Earth are being utilized. Agrarian productivity, harvesting of timber, exploitation of new oil and mineral fields, urban and rural growth, control of water resources, and other large scale interactions of man with his environment will have to be observed, monitored, controlled, and even actively managed in the future if we hope to maintain a decent living standard for large portions of mankind. Comprehensive surveying on a global scale, with very quick access to results, will be possible from orbiting stations. Skylab will help us develop sensors for the observations and techniques for data evaluation and distribution which, as tools for worldwide management, will be indispensable for the orderly growth of mankind.

• Advance scientific knowledge of the Sun and stars.

Ancient astronomers used the Sun's motion to predict the seasons and tell the best times for planting and harvesting. Their successors, modern solar astronomers, seek to understand and explain the remarkable phenomena within and around the Sun itself. While this is scientific research in its purest form, there is also a strong awareness that better knowledge of solar processes may lead the way to new means for generating and controlling energy for use on Earth.

The Skylab solar investigations address that area of interest and also the problem of explaining mechanisms by which solar events affect the Earth, particularly by the streams of high energy particles associated with solar flares. For example, it is assumed that these flares are responsible for the auroras and the associated disruption of ionospheric radio transmission. Since it is known that sunspot activity correlates with variations of temperature and density in the upper atmosphere, it is conceivable that the injection of energy in the upper atmosphere by solar particles may trigger worldwide weather phenomena as well.

The laboratory in orbit offers a unique opportunity to observe phenomena in the upper atmosphere, on the Sun, on other celestial bodies, and in the space between them, because Skylab will not be surrounded by the atmospheric filter that severely limits observational capabilities from the surface of the Earth.

- **Study the processing of materials under weightlessness.**

Society's demands for more and greater technological capabilities frequently lead to the limits imposed by existing materials. Fabrication of some of these materials is impeded by the effects of Earth gravity. Examples include large single crystals as needed in semiconductor technology, alloys containing components of widely differing densities, and superconducting materials with three or four dissimilar components. Exploration of the possibilities for producing such substances in an orbiting laboratory could lead to future large-scale zero-gravity production facilities.

- **Better understand manned space flight capabilities and basic biomedical processes.**

Certain trends and directions of future space flight activities have become evident. Man will use orbiting spacecraft because of their location above the atmosphere, because of their state of weightlessness, and because of their ability to see large portions of the Earth at one time. It is thinkable that future spacecraft could also be used for production plants whose output of heat and of polluting materials must be kept outside the Earth's atmosphere. It is also possible that future spacecraft will be needed to convert solar energy into useful electric energy for transmission to Earth. Undoubtedly, there will be other uses of orbiting satellites of which we have no clear idea yet, perhaps for therapeutical activities. For all these applications, there is a need to know the capabilities, limitations, and usefulness of man to live and work in space, and to act as a scientist, an engineer, a technician, an observer, a repairman, an evaluator, a medical doctor, a routine worker, a cook, a pilot and navigator, an explorer, a researcher, and simply as a crew member. Will the human body, which has been accommodated to the gravity field of the Earth ever since it has existed, readily adapt to life under weightlessness? Which instruments should be fully automated, and which should be man-operated? There is also a need to learn how the data systems can compress the enormous amount of data expected from Earth sensors so that only a selected portion need be transmitted to Earth stations. Man may have to play an important role in this selection and compression of observational data.

An orbiting laboratory offers a continuous state of weightlessness which cannot be obtained on Earth. It is believed that gravity may possibly be of influence in a number of biological and medical processes, such as the germination of seeds, the cleavage of cells, the growth of certain tissues, the regulation of metabolic processes, the adaptation of acceleration sensors, the control of cardiovascular functions, and perhaps the functioning of time rhythms. Studying these processes under weightlessness will help us understand their basic mechanism and some of the fundamental laws which govern live organisms.

In many respects, experience gained with Skylab will have a decisive influence on the further structure and conduct of space flight.

2. SKYLAB ELEMENTS

Fig. 7 shows Skylab in orbit. Its largest element is the Orbital Workshop (OWS), a cylindrical container of 15 meters (48 ft) length and 6.5 m (22 ft) diameter. The basic structure of this OWS is the third stage, or S–IVB stage, of the Saturn V which served as launch vehicle in the Apollo Program.

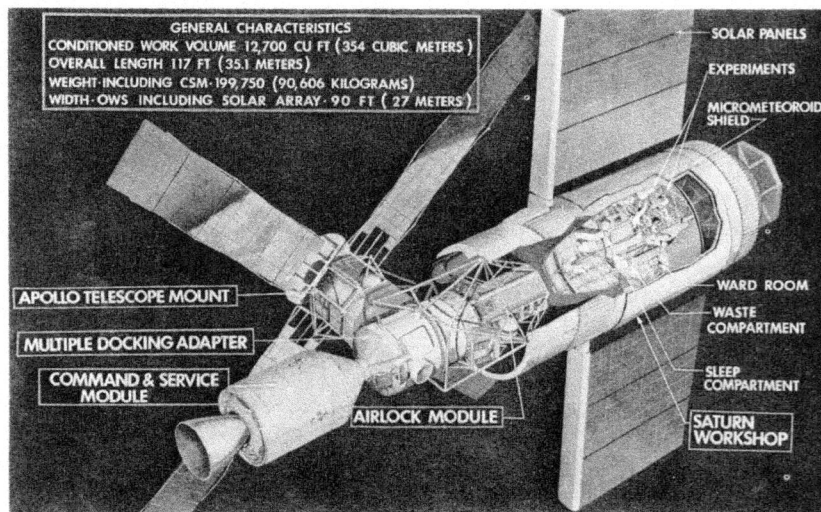

GENERAL CHARACTERISTICS
CONDITIONED WORK VOLUME 12,700 CU FT (354 CUBIC METERS)
OVERALL LENGTH 117 FT (35.1 METERS)
WEIGHT-INCLUDING CSM-199,750 (90,606 KILOGRAMS)
WIDTH-OWS INCLUDING SOLAR ARRAY · 90 FT (27 METERS)

SOLAR PANELS
EXPERIMENTS
MICROMETEOROID SHIELD

APOLLO TELESCOPE MOUNT
MULTIPLE DOCKING ADAPTER
COMMAND & SERVICE MODULE
AIRLOCK MODULE

WARD ROOM
WASTE COMPARTMENT
SLEEP COMPARTMENT
SATURN WORKSHOP

FIGURE 7.—Skylab in orbit.

This stage has been modified internally to work as a large orbiting space capsule rather than a propulsive stage. In designing the OWS as a place where a three-man crew can live and work for periods up to eight weeks, emphasis had to be shifted from the spartan austerity of earlier spacecraft to roomier, more comfortable accommodations.

Crew members will spend most of their time in the OWS, conducting experiments, making observations, eating, sleeping, and attending to their personal needs and wants.

Two further important elements are the Airlock Module (AM) (Fig. 8) and the Multiple Docking Adapter (MDA), (Fig. 9). The AM enables crew members to make brief excursions outside Skylab as required for experiment support. Separated from the Workshop and the MDA by doors, the AM can be evacuated for egress or ingress of a space-suited astronaut through a side hatch. Oxygen and nitrogen storage tanks needed for Skylab's life support system are mounted on the external trusswork of the AM. Major components in the AM include Skylab's electric power control and distribution station, environmental control system, communications system, and data handling and recording systems. The AM may be called the "nerve center" of Skylab.

14

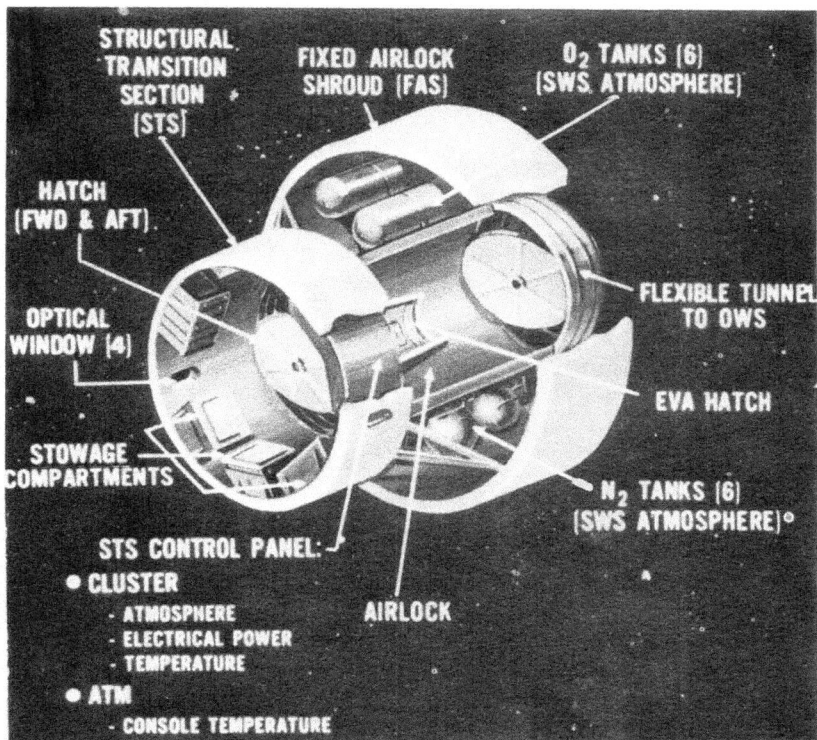

STRUCTURAL TRANSITION SECTION (STS)

FIXED AIRLOCK SHROUD (FAS)

O₂ TANKS (6) (SWS ATMOSPHERE)

HATCH (FWD & AFT)

OPTICAL WINDOW (4)

FLEXIBLE TUNNEL TO OWS

EVA HATCH

STOWAGE COMPARTMENTS

N₂ TANKS (6) (SWS ATMOSPHERE)°

STS CONTROL PANEL:

● CLUSTER
   - ATMOSPHERE
   - ELECTRICAL POWER
   - TEMPERATURE

AIRLOCK

● ATM
   - CONSOLE TEMPERATURE

FIGURE 8.—Airlock Module (AM).

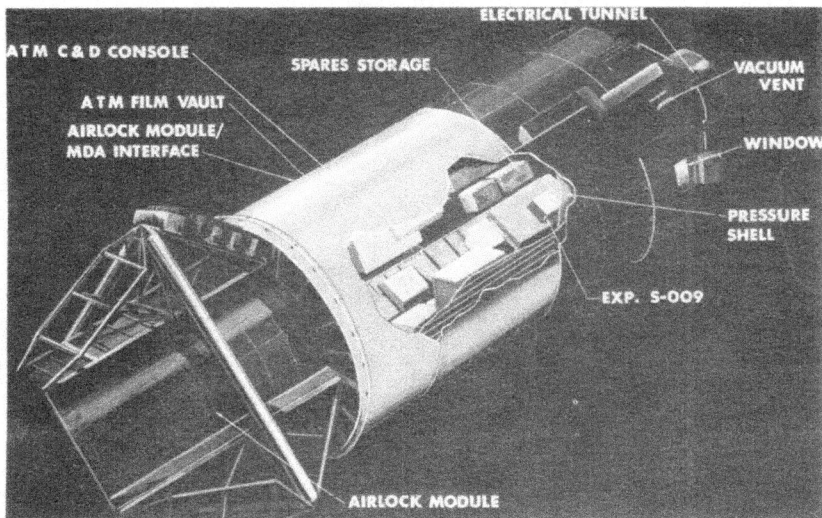

ATM C & D CONSOLE

SPARES STORAGE

ELECTRICAL TUNNEL

VACUUM VENT

ATM FILM VAULT

AIRLOCK MODULE/ MDA INTERFACE

WINDOW

PRESSURE SHELL

EXP. S-009

AIRLOCK MODULE

FIGURE 9.—Multiple Docking Adapter (MDA).

15

Forward of the AM is the Multiple Docking Adapter (MDA) which provides docking facilities for the Command and Service Module (CSM). Under normal operations, the CSM will dock at the forward end of the MDA (Fig. 10). In case of an emergency, a side docking port can be used. Two CSM's could even dock simultaneously, if the need should arise. Besides providing mechanical support for the Apollo Telescope Mount, the MDA also accommodates several experiment systems, among them the Earth Resources Experiment Package (EREP), the materials processing facility, and the control and display console needed for ATM solar astronomy studies.

The CSM that serves as Skylab's transportation system and communications center is an example of how Apollo-proven systems are used on Skylab. Each of the three CSM's to be used with Skylab has been modified to meet specific requirements. A rescue kit has been developed for use if needed which will enable a CSM to carry as many as five astronauts back to Earth.

The external experiment assembly known as the Apollo Telescope Mount (ATM, Fig. 11), contains eight complex astronomical instruments designed to observe the Sun over a wide spectrum from visible light to X-rays. These instruments, together with auxiliary equipment, have been integrated into a common system as shown in Fig. 12. During launch, the ATM module will be in an axial position, as shown in Fig. 13. As soon as Skylab reaches orbit, the ATM and its windmill-like solar cell array that converts solar energy to electric power for Skylab will be deployed, and Skylab will be oriented so that ATM instruments and solar cell panels face the Sun.

The ATM control moment gyro system provides the primary attitude control for Skylab. An independent pointing control system of limited angular freedom will allow precise pointing of the solar astronomy experiments.

FIGURE 10.—The Command and Service Module (CSM) docked with Skylab. An astronaut is performing Extravehicular Activity (EVA) at the front end of the Apollo Telescope Mount (ATM).

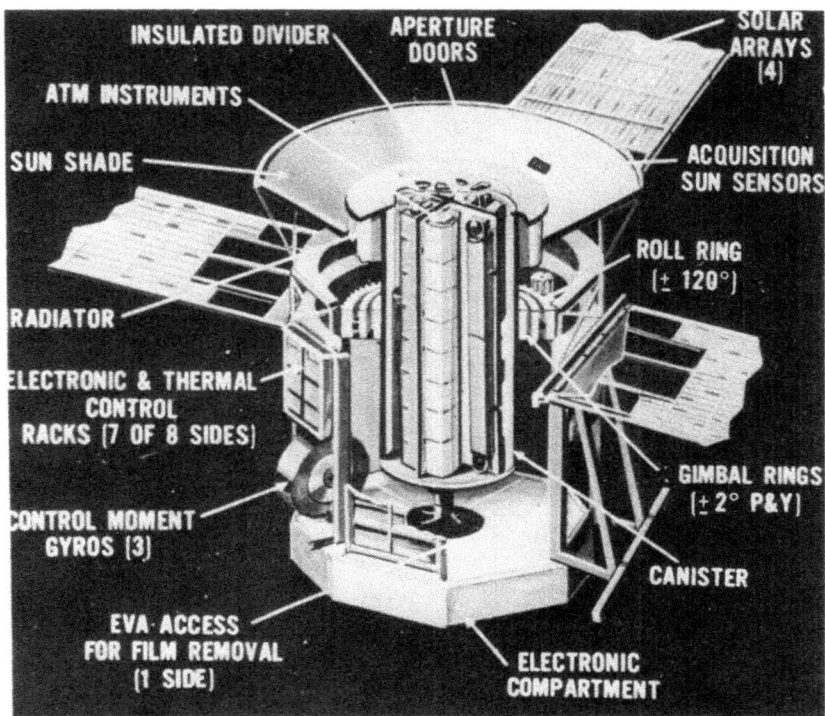

FIGURE 11.—The Apollo Telescope Mount (ATM).

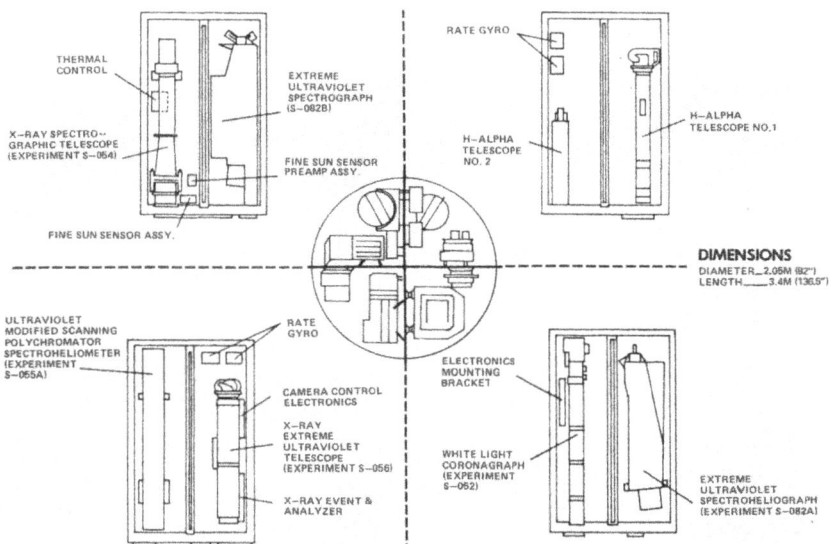

FIGURE 12.—Cross sections through Apollo Telescope Mount (ATM), showing individual telescopes.

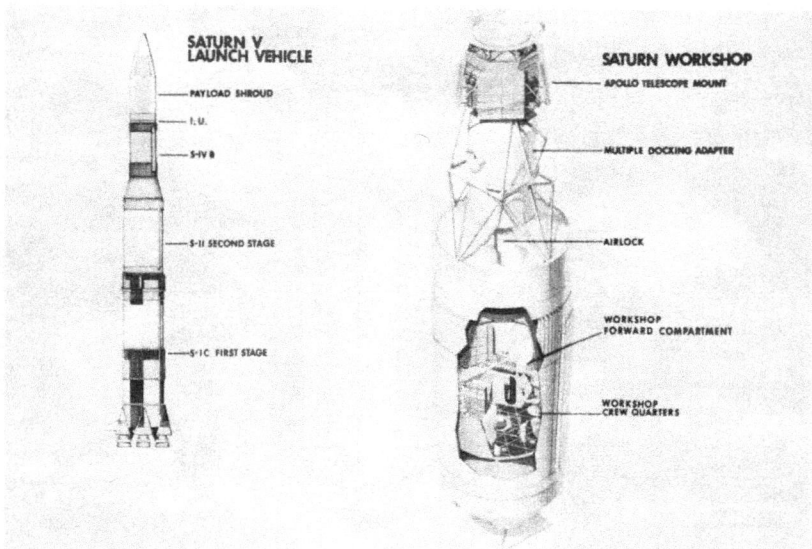

FIGURE 13.—Apollo Telescope Mount (ATM) and other Skylab components in launch configuration on top of Saturn V.

3. MISSION PLAN

Skylab, consisting of Orbital Workshop, Airlock Module, Multiple Docking Adapter, and Apollo Telescope Mount, will be launched as one unit with a two-stage Saturn V launch vehicle (Fig. 14). After arrival in orbit,

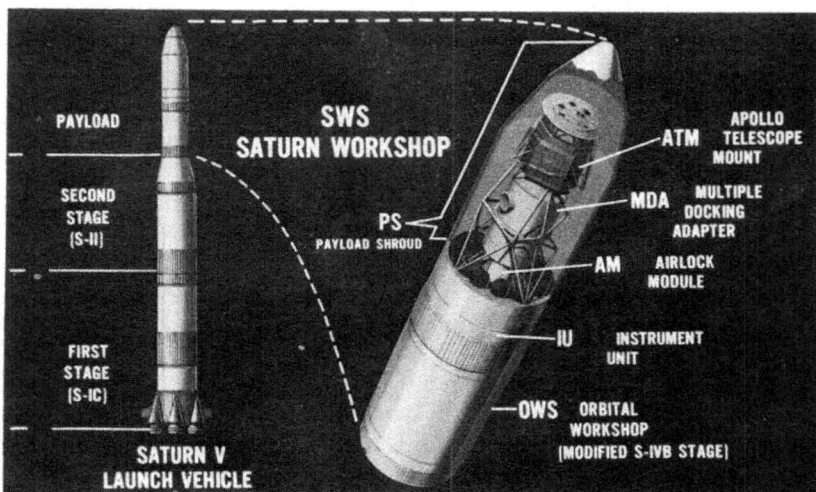

FIGURE 14.—Skylab, packaged for launch on top of Saturn V.

18

and after the maneuver of rotation and deployment, all systems will be confirmed as operationally ready, by automated or by remotely controlled actions. One day after Skylab launch, the first crew of three astronauts will be launched in the Command and Service Module by a Saturn IB launch vehicle. After docking with Skylab, the crew will check out and fully activate all systems; Skylab will then be ready for its first operational period of 28 days (Fig. 15). At the end of this period, the crew will return to Earth with the CSM, and Skylab will continue some of its activities over a two-month period of unmanned operation. Three months after the first crew launch, the second three-man crew will be launched with the second Saturn IB, this time for a 56-day period of manned operation. After return of the second crew to Earth, Skylab will operate in its unmanned mode for a month. The third three-man crew will then be launched with the third Saturn IB, again for a 56-day period in orbit. Although Skylab will remain in orbit for years after the third crew has returned, unmanned operation of Skylab will continue only for a short time because some of the Skylab systems will reach the end of their design lifetimes.

Total Skylab mission activities will cover a period of roughly eight months. The distribution of manned and unmanned phases is shown in Fig. 16. The initial step in the Skylab mission will be the launch of a two-stage Saturn V booster, consisting of the S–1C first stage and the S–II second stage, from Launch Complex 39A at the Kennedy Space Center in Florida (Fig. 17). Its payload will be the unmanned Skylab, which consists of the Orbital Workshop (OWS), the Airlock Module (AM), the Multiple Docking Adapter (MDA), the Apollo Telescope Mount (ATM), and an Instrument Unit (IU).

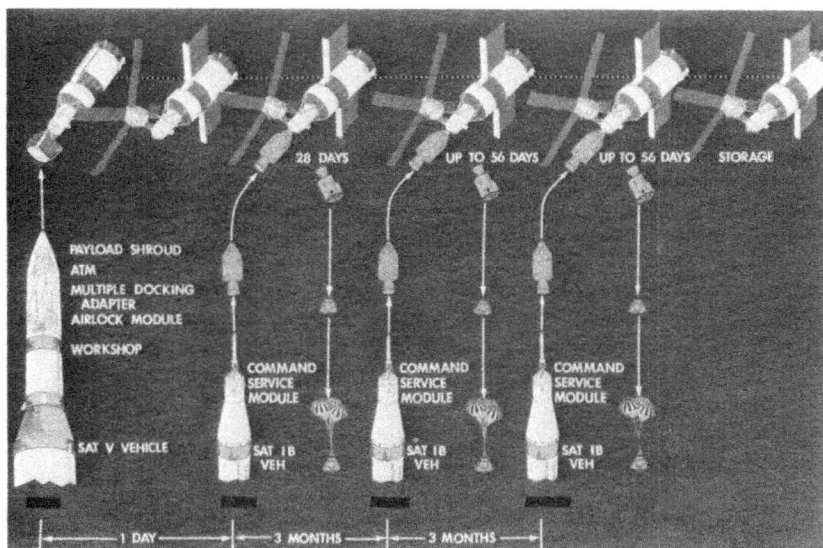

FIGURE 15.—Launch sequence of Skylab and three manned Command and Service Modules (CSM).

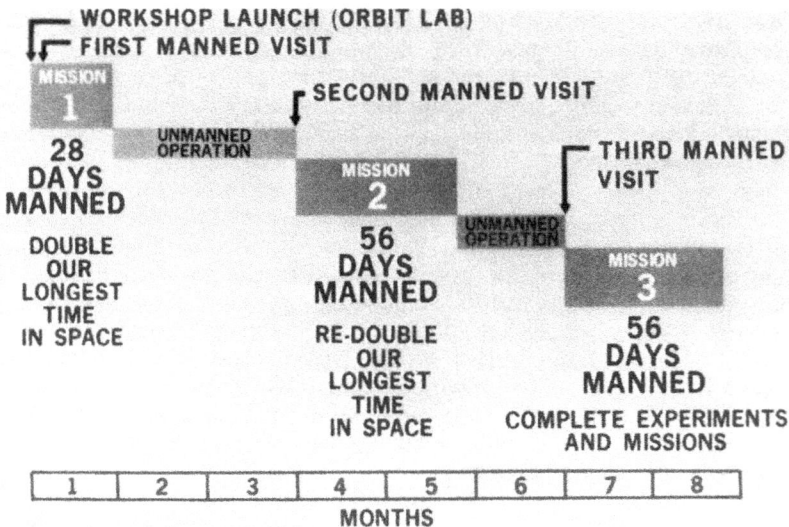

FIGURE 16.—Manned and unmanned phases of the Skylab mission.

Skylab will be inserted into a near-circular orbit at an altitude of 432 km (268 statute miles or 234 nautical miles) with an orbital inclination of 50 degrees to the Earth's equator (Fig. 18). Fig. 19 shows the Skylab, with shroud in launch configuration shortly after separation from the second stage of the Saturn V. After launch, the following events will occur (Fig. 20):

Jettison of the Payload Shroud,
Deployment of the Apollo Telescope Mount,
Extension of the ATM solar arrays,
Rotation of the vehicle until the ATM solar arrays point at the sun,
Extension of the Orbital Workshop solar arrays,

Pressurization of the habitable areas to 34,000 Nm⁻² (one-third of an atmosphere, or 5 psi).[1] The breathing gas consists of 74 percent oxygen and 26 percent nitrogen (by volume).

About 24 hours after the Skylab launch, the manned CSM will be launched with the three astronauts who will occupy Skylab for 28 days. This launch, taking off from Launch Complex 39B, will use a Saturn IB to boost the Command and Service Module (Fig. 21) with three astronauts into an interim elliptical orbit with a perigee of 149 km (93 statute miles or 80 nautical miles), and an apogee of 223 km (138 statute miles or 120 nautical miles). From this interim orbit, the CSM will use the Service Module propulsion system to transfer to the Skylab orbit (432 km or 268 statute miles or 234 nautical miles) in order to rendezvous with the spacecraft and to dock at the end port of the Multiple Docking Adapter (Fig. 22). This docking will complete the Skylab cluster.

---

[1] 1 Newton per square meter=1 Nm⁻²; 100,000 Nm⁻²=.987 atmosphere=14.5 pounds per square inch.

FIGURE 17.—Saturn V with Skylab on the launch tower.

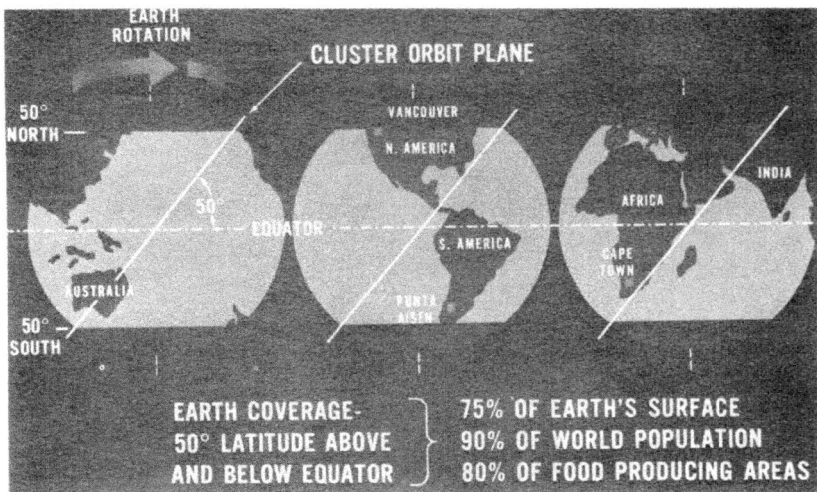

FIGURE 18.—Coverage of a broad region on the earth's surface by Skylab.

FIGURE 19.—Separation of the Skylab (with shroud) from the second stage of the Saturn V vehicle.

The astronaut crew will enter and activate Skylab for its manned missions. Only the essential elements of communications, instrumentation, and thermal control systems of the CSM will remain in operation.

During the 28 days of the first manned mission, the astronaut crew on Skylab will conduct experiment programs and evaluate the habitability of Skylab (Fig. 23). It is planned to obtain data from all but a few experiments during this mission. At the end of the 28-day period, the astronaut crew will prepare the cluster for unmanned operation, transfer to the CSM, and separate from Skylab. A deceleration maneuver, exe-

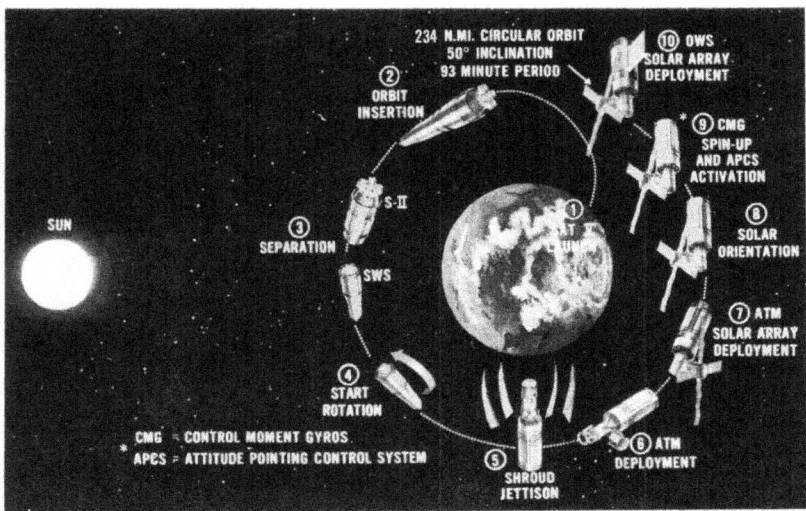

FIGURE 20.—Launch and deployment phases of Skylab during the first mission day.

cuted by firing the Service Module engine, will cause the CSM to lose velocity and reenter the atmosphere. Shortly before reentry, the Command Module will separate from the Service Module and a little later will descend by parachute to the Pacific recovery area (Fig. 24).

The second manned mission will start with another Saturn IB launch from Complex 39B approximately 60 days after return of the first crew. Orbit insertion, rendezvous, and docking procedures will follow the pattern of the previous flight. Activities performed by the crew after transfer to Skylab will be similar to those in the previous mission; however, more emphasis will be placed on solar astronomy and Earth resources observations. The mission duration will be increased to 56 days, with recovery again in the Pacific.

The third manned mission again will be launched from Launch Complex 39B about 30 days after the second crew has returned. In this mission, also of 56 days, the Skylab experiment program will be continued, and additional statistical data will be obtained on the crew's adaptability and performance. Recovery of the Command Module with crew and data will occur in the mid-Pacific area.

Owing to its orbital inclination of 50 degrees, the trajectory of Skylab sweeps over a large portion of the Earth's surface. Crew and instruments on Skylab will be able to see about 75 percent of the Earth, including all of Africa, China, and Australia, almost all of South America, most of North America, and much of Europe and northern Asia. The pattern of ground tracks, illustrated in Figs. 25 and 26, will repeat itself every five days. At least one ground tracking station will be overflown during each orbit. However, there are periods up to about an hour's duration on each orbit during which Skylab will not be within radio or telemetry reach of any station. During

23

FIGURE 21.—Saturn IB with the manned Command and Service Module (CSM) on the launch tower.

these periods, voice and data will be recorded on tape for quick replay and transmission while Skylab is in contact with one of the ground stations. A more detailed description of the data and communication system will be given in Chapter IV.2.c.

Unique in the Skylab Program is the ability to rescue astronauts in space, if the need should arise. This rescue capability exists because the Orbital Workshop offers long-duration life support in Earth orbit. The Skylab rescue capability is described in the next section.

4. RESCUE CAPABILITIES

The Skylab Program includes the capability to rescue astronauts under certain circumstances.

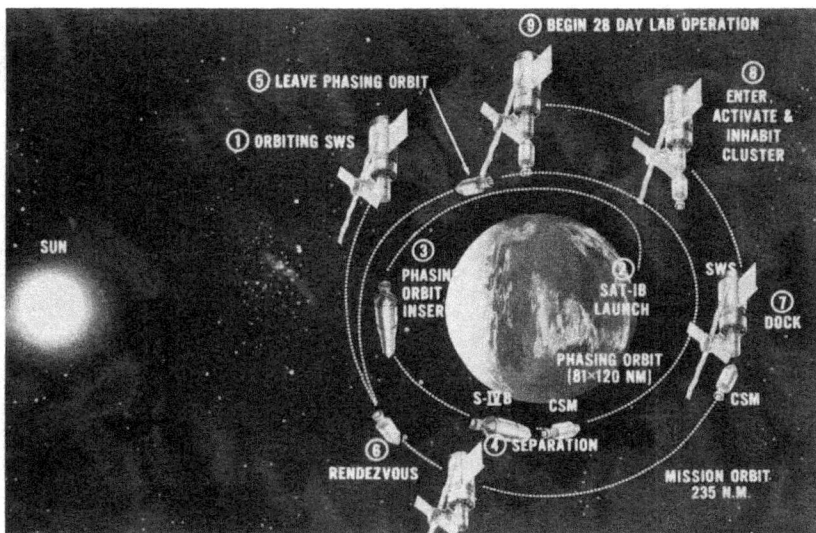

FIGURE 22.—Phases of launch rendezvous and desking of the Command and Service Module (CSM) with the orbiting Skylab during the second day of the Skylab mission.

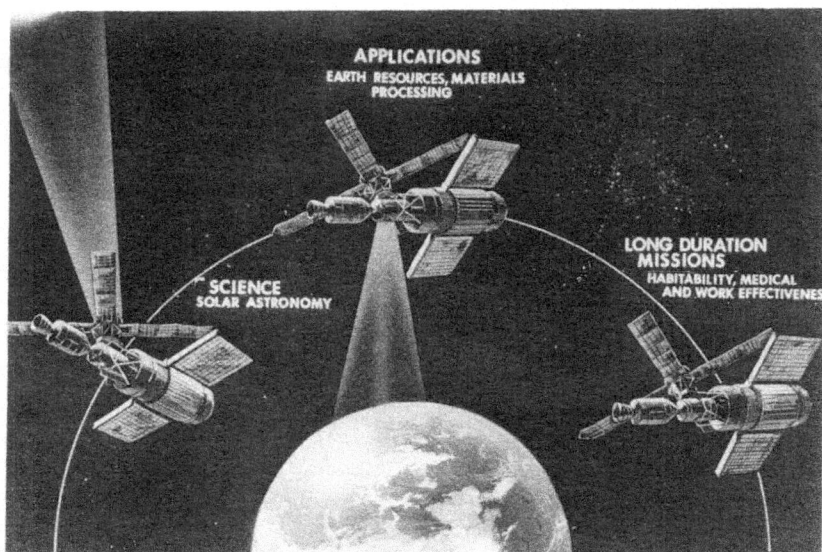

FIGURE 23.—Major program phases of Skylab during its entire mission.

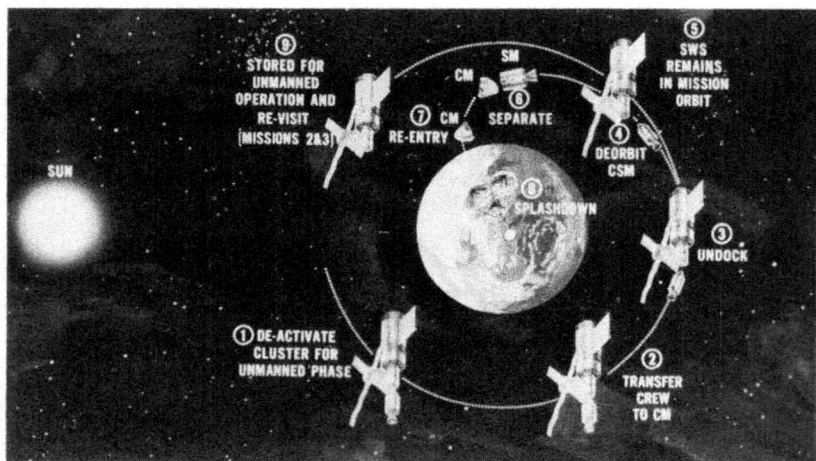

Figure 24.—Undocking, deorbiting, separation, reentry, and splashdown of the Command and Service Module (CSM) during the last day of the first manned mission.

During each of the three Skylab visits, the astronauts will be transported to and from the orbiting cluster in a modified Apollo Command and Service Module. After docking the Skylab activation, the CSM will be powered down, but it will remain available for life support and normal crew return. It always will be ready for quick occupation by the astronauts in the event of a serious failure in the cluster.

The Skylab cluster's ample supplies and long-duration life support capability make it possible to rescue the astronauts in the event that the CSM which brought them up to the cluster becomes unuseable for recovery. Therefore, the only failures to be considered for rescue requirements are loss of CSM return capability or loss of accessibility to the CSM. In either of thse events, a second CSM would be launched with two men on board and with room for the three astronauts to be picked up in orbit. The rescue CSM would then return with five crew members.

How long the Skylab astronauts would have to wait for rescue would depend on the time during the mission schedule when the emergency develops. The waiting time could vary from 10 to 48 days.

The three manned launches in the Skylab Program will be about 90 days apart. After each of the first two manned launches, the next vehicle in normal preparation for launch would be used for rescue, if needed. After the third and final manned launch, the Skylab backup vehicle will be kept ready for possible use as a rescue spacecraft.

If the need for rescue arose on the first day of Skylab's occupancy or reoccupancy, present work schedules indicate that it would take 48 days for the launch crews to ready the rescue launch vehicle and spacecraft. This time includes 22 days which would be required to refurbish the launch tower at Pad B of Complex 39 following the previous launch. During this period, the specially-developed Command Module rescue kit would be installed, a task which would take only about eight hours.

TYPICAL SKYLAB GROUND TRACK
REVOLUTIONS 57 THRU 63

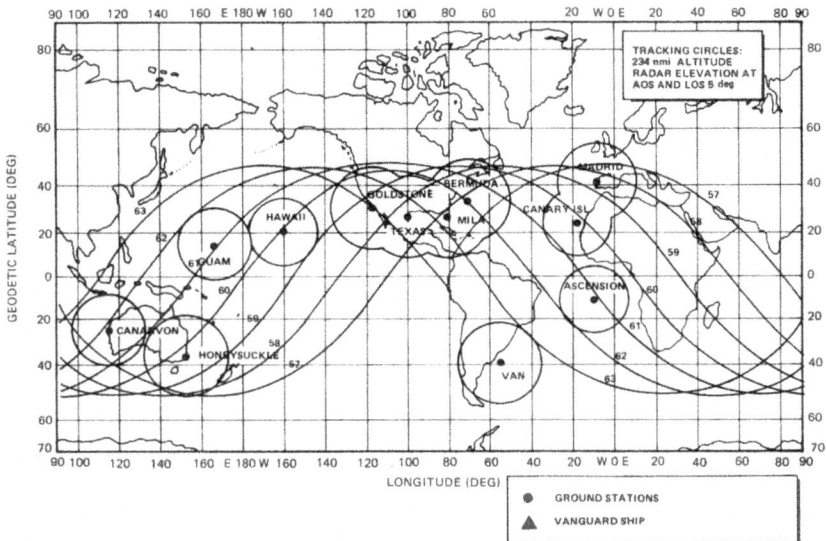

TYPICAL SKYLAB GROUND TRACK
REVOLUTIONS 64 THRU 70

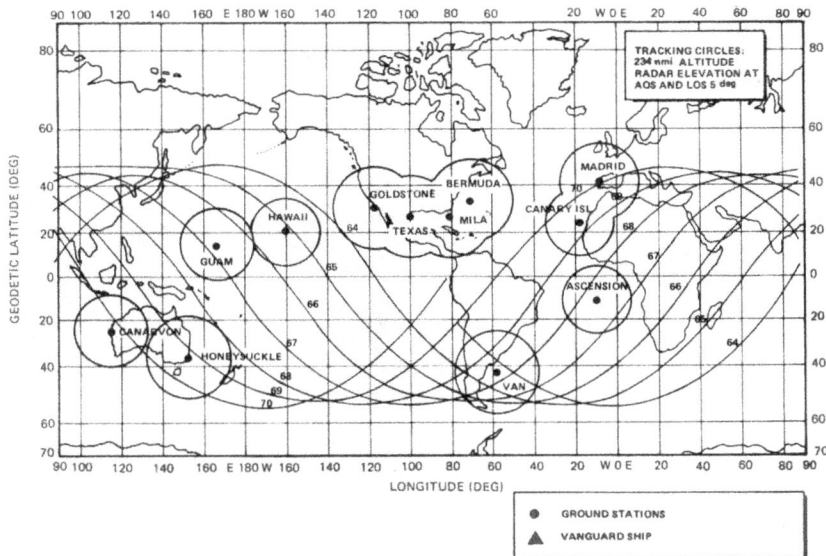

FIGURES 25 & 26.—Ground tracks of Skylab during selected days of its mission.

The later during a mission the need for rescue arises, the sooner the rescue vehicle could be made ready for launch. Launch readiness time will be reduced to 28 days at the end of the first manned visit and to 10 days at the end of the third mission.

To convert the standard CSM to a rescue vehicle, the storage lockers would be removed and replaced with two crew couches to accommodate a total of five crewmen. (Fig. 27.)

Prior to rescue, the stranded Skylab crew members would don pressure suits and enter the Multiple Docking Adapter, seal it off from the remainder of the cluster, and depressurize it. Then they would install a special spring-loaded device to separate the disabled CSM from the axial port of the MDA at sufficient velocity to move it out of the way of the arriving rescue CSM. However, this is not absolutely necessary. The arriving CSM could also dock at the radial (side) docking port of the MDA. In this position, which is a contingency mode, limited but sufficient stay time would be available for full rescue operations.

Providing rescue modes for all conceivable emergency situations would require instantaneous rescue response, a capability not feasible with present space vehicles because of elaborate launch preparations. Faster response must await a new generation of space transportation systems such as the Space Shuttle. However, the planned rescue techniques for Skylab will cover the most likely emergency situations, adding a new dimension of flexibility and safety to manned space flight.

5. CREW ACTIVITIES

During each of the manned periods, a typical crew day consists of eight hours per crewman for experiment activities; eight hours for eating, personal

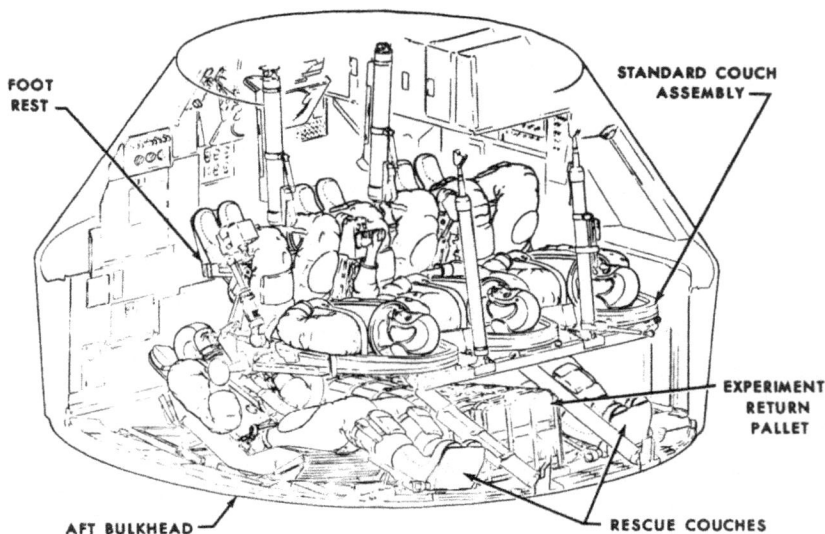

FIGURE 27.—Command Module, modified for rescue operations by removal of storage lockers, occupied by five astronauts.

hygiene, systems housekeeping, mission planning, and off-duty functions; and eight hours for sleeping (Fig. 28 and 29). As a rule, the three crewmen will sleep during the same period of eight hours when all the functions on board Skylab will be automatic. During eating and off-duty periods, crewmen may slightly shift their activities in such a way that one member of the crew is always available for Sun viewing at the ATM console. Planned schedules for operation of the experiments during the first manned mission are illustrated in Fig. 30.

Some experiments will be operative continuously, others will operate on several or all days for limited periods, and still others will have only one short period of operation during a mission.

Upon rendezvous and docking with Skylab, the astronaut crews will be primarily concerned with the following activities:

Activating, operating, and monitoring Skylab systems;
Conducting numerous experiments in the medical, scientific, engineering, technology, and earth resources categories;

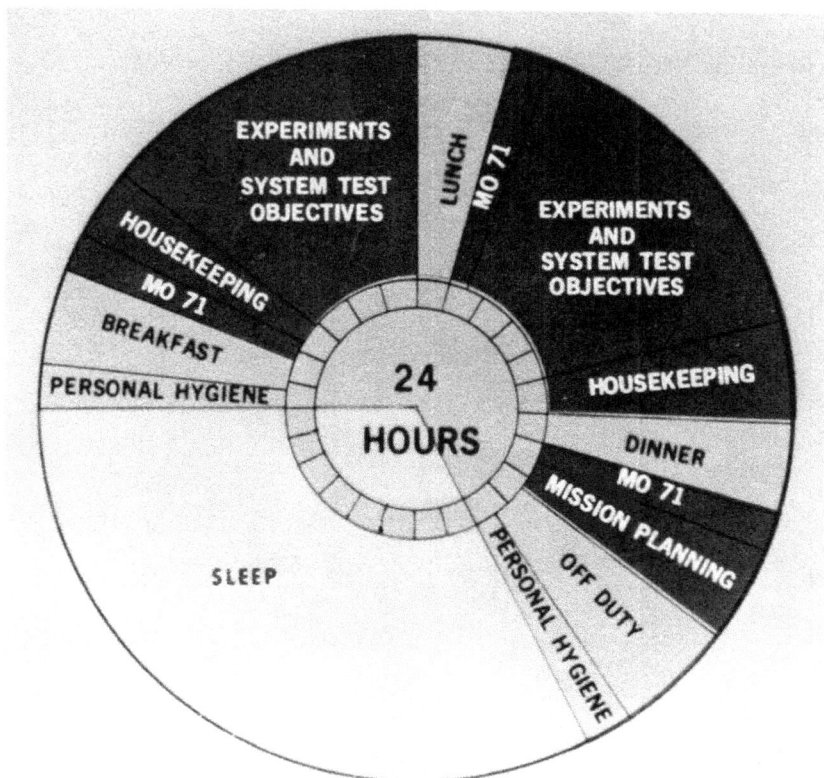

FIGURE 28.—Astronaut activities during a typical 24-hour period.

Conducting Extravehicular Activities (EVA) ;
Personal activities.

These activities are discussed in detail in following paragraphs.

Activating, Operating, and Monitoring Skylab Systems

Individual actions of the astronauts during the activation period of Multiple Docking Adapter, Airlock Module, and Workshop after the CSM has docked at the axial port of the MDA are listed on the Manned Activation Schedule, Table 3. A similar schedule exists for deactivation of the cluster shortly before the end of the 28-day period of occupation. The activation and deactivation schedules for the other two manned periods resemble those for the first manned period.

An important function of the astronaut crew is to monitor the following subsystems and their functions on Skylab: attitude control, electric power, environment and thermal control, and instrumentation and communications.

These subsystems are discussed in detail in Chapter IV-2.

Operating the Experiments

The main floor of the Orbital Workshop includes an experimental work area immediately adjacent to the wardroom (Fig. 32). Primary medical experiments to assess the manner in which the astronauts are adjusting to spaceflight will be conducted in this area. To support these experiments,

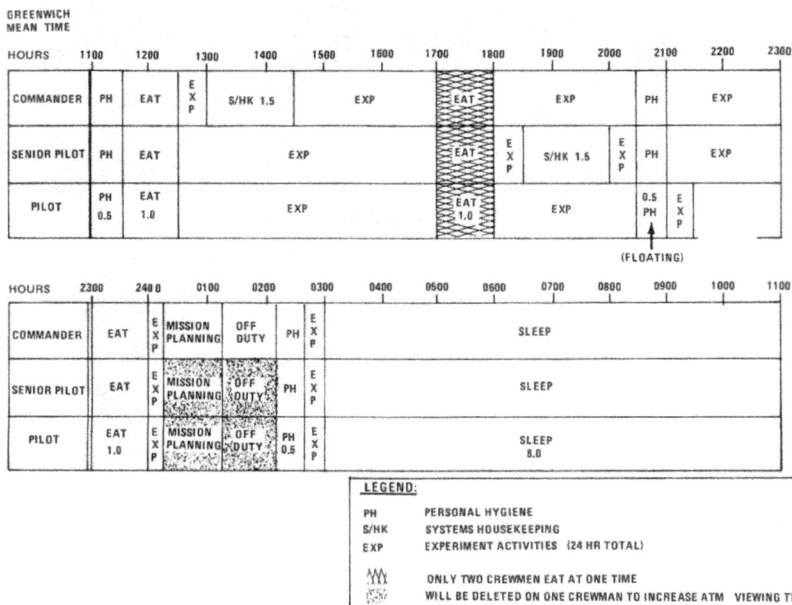

FIGURE 29.—Schedule of activities for the three astronauts during a typical day of the Skylab mission.

TABLE 3.—*Manned Activation Schedule: First Manned Period*

Time—day, hour, minute	Action
01:07:29........	Dock CSM to axial port of MDA (reference).
	Activate CM/MDA docking tunnel (30 min):
	Turn on CM tunnel lights.
	Verify integrity of CM/MDA docking tunnel pressure seal.
	Open CM forward hatch pressure equalization valve.
	Verify CM docking tunnel pressure.
	Open & stow CM forward hatch.
	Verify that docking latches are secure.
	Change attitude & pointing control system control gains to docked configuration.
	Remove & stow docking probe drogue.
01:07:59........	Crew Functions (eating, sleeping, personal hygiene, etc—11 hr 30 min).
01:19:29........	Activate MDA/AM (2 hr):
	Verify MDA/AM pressure/temperature.
	Open & secure MDA hatch.
	Close MDA hatch pressure equalization valve/pressure cap.
	Turn on MDA interior lights.
	Cap MDA vent line.
	Turn on molecular sieve fans; A primary, B secondary.
	Perform MDA initial entry inspection.
	Activate MDA window frame heater & thermally condition film.
	Turn on STS forward & aft floodlights.
	Perform STS/forward AM initial entry inspection.
	Turn on MDA/AM and MDA area fans.
	Activate STS controls and displays console.
	Turn on AM duct fan.
	Connect CM/MDA signal/electrical Power system umbilical in MDA tunnel.
	Transfer single-point ground from AM to CSM.
	Activate and check out caution and warning system.
	Complete CM/SWS signal transfer circuits.
	Activate and check out MDA/AM speaker intercoms.
	Switch AM coolant system from ground to manual control.
	Install MDA environmental control system air interchange duct in CM.
	Turn on CM duct fans.
	Switch AM/MDA heater from ground to manual control.
	Turn off OWS radiant heaters.
	Activate MDA/AM environmental control system.
	Activate ATM controls and displays console/earth resources experiment package cooling system.
	Begin initial activation of ATM controls and displays console.
	Verify matching of CM/SWS power systems.
	Begin final activation of ATM controls and displays console.
	Verify that differential pressure across forward AM hatch is 0.0 psid.
	Open forward AM hatch.
	Turn on AM lock compartment lights.
	Engage EVA hatch handle retainer pin.
	Verify that differential pressure across aft AM hatch is 0.0 psid.
	Open Aft AM hatch.
	Turn on Aft AM & OWS initial entry lights.
	Position OWS heat exchanger fan switches to OWS position.

Table 3.—*Manned Activation Schedule: First Manned Period*—Continued

Time—day, hour, minute	Action
01:21:29........	Activate OWS (1 hr 30 min):
	Verify OWS pressure.
	Open OWS hatch pressure equalization valve.
	Open OWS quick-opening hatch and lock in open position.
	Cap OWS pneumatic and solenoid vent valve lines.
	Perform OWS initial entry inspection.
	Activate and check out speaker intercom station.
	Connect flexible duct from AM to OWS.
	Install OWS fireman's pole.
	Divert air flow from MDA to OWS.
	Activate OWS duct fans.
	Activate OWS controls and displays console.
	Check out OWS caution and warning system.
	Activate OWS thermal control system.
	Release fire extinguisher from launch restraints.
	Deactivate humidity control in OWS film vault.
	Turn off OWS initial entry lights.
	Transfer equipment from CM to OWS.
01:22:59........	Begin activation of OWS habitation support system (30 min).
	Inspect wardroom and sleep/waste management compartments.
	Activate OWS wardroom window heater .
	Activate waste management system.
	Activate food management system.
01:23:29........	Crew functions (lunch, personal hygiene—1 hr 30 min).
02:00:59........	Complete activation of OWS habitation support system (1 hr 5 min).
	Activate water system.
	Check out trash disposal airlock.
	Verify that contents are in stowage containers.
	Activate and check out vacuum cleaner.
	Permanently stow docking probe and drogue in MDA.
02:02:04........	Begin initial setup and checkout of experiments (2 hr 25 min).
02:04:29........	Crew functions (eating, personal hygiene—1 hr).
02:05:29........	Complete experiment setup/checkout (3 hr).
02:08:29........	Crew functions (personal hygiene thru 8 hr sleep—9 hr).
02:17:29........	End manned activation (begin first on-orbit typical crew day).

Activity	01	02	03	04	05	06	07	08	09	10	11	12	13	14	15	16	17	18	19	20	21	22	23	24	25	26	27	28	29	Total Mission Time (hr:min)
Activation	X	X																												7:30
Off Duty							X							X								X								15:00
EVA																										X				7:00
ATM			X	X	X	X		X	X	X	X	X	X	X		X	X	X	X	X	X	X	X	X		X	X			190:30
D008																			X		X									3:04
EPEP				X					X	X	X	X		X		X		X			X	X	X							24:15
M071/73	X	X	X	X	X	X	X	X	X	X	X	X	X	X	X	X	X	X	X	X	X	X	X	X	X	X	X	X	X	117:04
D024																										X				Included in EVA
M074			X														X											X		3:00
M172			X													X												X		1:30
M092/93										X						X							X							16:30
M092/171			X	X			X							X					X						X		X	X		35:42
M131					X	X			X				X		X					X	X			X						16:14
M133				X	X	X		X			X			X			X		X			X		X	X	X				3:15
M151	←								Included in Other Experiments																				→	
M415	←								No Astronaut Activity																				→	
M487					X			X				X	X			X	X	X												5:25
M509					X									X					X						X					8:48
M512					X				X			X	X								X									7:06
M516	←								Included in Other Experiments																				→	
S009			X	-	-	-	-	-	-	-	-	-	-	-	-	-	-	-	-	-	-	-	-	-	X					:30
S015	X	X	X	X	X	X	X	X	X	X																				2:24
S019							X			X	X	X															X	X		9:15
S073/T027			X	X	X								X	X				X	X		X			X			X			27:37
S149															X	X											X		-	2:10
S190B									X	X				X			X			X				X	X					8:37
T003			X	X	X	X	X	X	X	X	X	X	X	X	X	X	X	X	X	X	X	X	X	X	X	X	X	X	X	11:56
S063															X							X	X							4:25
T027			X	-	-	-	-	-	X																					1:15
Reentry Simulation																						X								4:00
Deactivation																												X	X	8:45

Note: - - - indicates no crew time required during experiment operation.

FIGURE 30.—Schedule showing time allocations for Skylab experiments during first manned mission.

the astronauts will utilize such major equipment as the Lower Body Negative Pressure Device; Ergometer and Vectorcardiogram for cardiovascular experiments; Metabolic Analyzer for the pulmonary experiment, which also uses the Ergometer; the Rotating Chair for the vestibular (neurophysiology) experiment; and the Experiment Support System which supports several medical experiments with displays, data handling, controls, and power supplies. These experiments are further described in Chapter V–3.

Scientific experiments will feature observations of the Sun with the Apollo Telescope Mount (ATM). This solar observatory has several telescopes for studies of the Sun over a wide range of the spectrum. Astronauts will, through onboard displays, visually scan the Sun to locate targets of scientific interest (Fig. 33). They will assist in the alignment and calibration of the instruments, point them to the appropriate targets, make judgments of operating modes, and generally conduct a comprehensive program of solar investigation. The crew is to compensate as much as possible for any equipment failure to preserve the value of the ATM's scientific returns. Crew members will retrieve, through extravehicular activity, the photographic films on which solar data will be recorded.

Skylab's Earth Resources Observations will permit simultaneous remote sensing from orbital altitudes in the visible, infrared, and microwave spectral regions. Data thus obtained will be correlated with information obtained simultaneously about some of the same sites from aircraft and by ground

FIGURE 32.—Work area for medical experiments within Orbiting Workshop (OWS).

FIGURE 33.—Control and display console for astronaut operation of the Apollo Telescope Mount.

measurements. The astronauts will acquire preselected primary or alternate targets, operate equipment associated with the Earth Resources Experiment Package (EREP) such as sensors and cameras, and supply and retrieve film from the Earth Terrain Camera. Another major function of the crew will involve coordination with ground-based activities and Mission Control to update EREP operations.

The technological experiments will also be conducted in the MDA. In most instances, the involvement of man is a critical element in the successful accomplishment of these technological experiments. Data from these studies are extremely important in the development of future space projects for the conduct of scientific experimentation.

Extravehicular Activity (EVA)

The Extravehicular Activity performed by the astronauts is concerned primarily with changing cameras and film magazines on the solar telescopes, but samples also will be retrieved from the DO24 experiment (Thermal Control Coatings, Chapter V–4–d and VIII). Each EVA excursion will last up to three hours from start of egress (leaving Skylab) to completion of ingress (entering Skylab). One EVA of three hours duration is planned during the first mission on Day 26 for the Thermal Control Coatings Experiment DO24 and for ATM film change. Three EVA's for ATM film change will be achieved during the second mission and two EVA's for ATM film removal during the third mission (Fig. 34).

Two crewmen will don their space suits for each EVA. One of them will execute the outside activities, while the other one will stand ready to help,

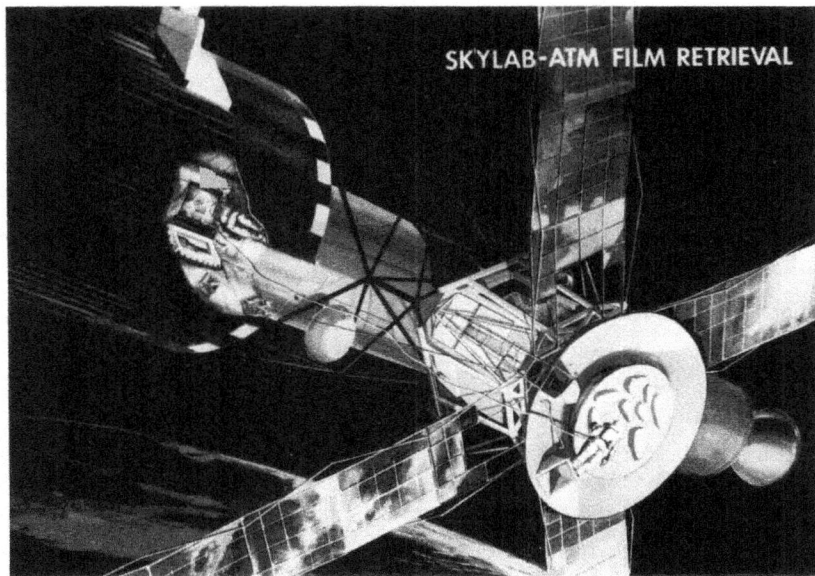

SKYLAB-ATM FILM RETRIEVAL

FIGURE 34.—Astronaut retrieving film during Extravehicular Activity (EVA).

if needed. The third astronaut will stay in the Airlock Module during EVA where he will perform monitoring and housekeeping activities.

Personal Activities

To provide entertainment and recreation for the astronaut crew members during their off-duty hours, the following items are furnished: audio equipment, playing cards, library, dart game, exercise equipment, binoculars, and balls. This equipment is described in the following paragraphs.

Audio Entertainment Equipment.—A permanently located tape player provides monaural or stereophonic playback of prerecorded tape cassettes through speakers or headsets. For individual use, headsets with head-phone plugs are available along with headset earpieces for connection to the tape player. Forty-eight cassettes of either stereophonic or monaural prerecorded material selected by the crews are provided.

Playing Cards Equipment.—Four decks of standard playing cards are provided along with five card deck retainers and five card retainers to permit card playing under zero gravity. The retainers hold the cards in place as a deck (card deck retainer), or individually for player use (card retainers). Cards are played normally at the food table with the table top in place. The retainers are held to the table top by magnets.

Library.—Approximately 36 crew-selected paperback books compose the library.

Dart Throwing Equipment.—Twelve darts and a dart board are stowed on Skylab. The board has Velcro hooks on the back for placement at any convenient Velcro location in the OWS. The target side of the dart board has Velcro pile superimposed on a standard target face. Each dart is a standard dart with Velcro hooks substituted for the pointed shaft.

Exercise Equipment.—A stationary bicycle exercise machine will be on board (Fig. 35). In addition, each crew member will be provided with an isometric exerciser for in-flight exercising. Six hand exercisers, shaped to fit the hand, are also used to maintain grip strength.

Binoculars.—A pair of center-focus binoculars is also available.

Game Balls.—Three plastic-covered foam balls are furnished for recreation.

In addition to recreational activities, personal activities include sleeping, eating, and personal hygiene. These topics are discussed in detail in Chapter IV–2–d on Crew Accommodations.

6. CREW TRAINING PROGRAM

Success of a manned mission largely depends upon the astronauts' ability to perform their assigned functions properly and effectively. In a mission as complex as Skylab, thorough crew training is essential so that the astronauts can perform their many experiments and their housekeeping duties within the limited time of their missions, and even can respond to emergencies if and when these should arise.

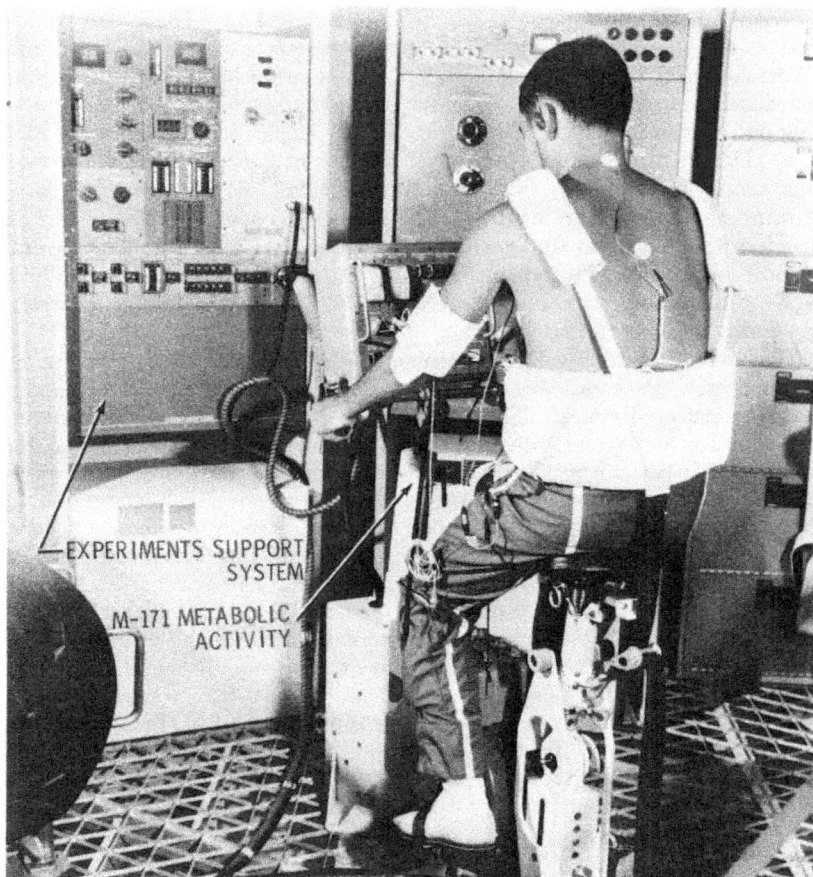

FIGURE 35.—Astronaut training on the wheelless bicycle (Ergometer) within Workshop.

Primary and backup crews for all Skylab missions have received identical training. If a need should arise for crew replacement, substitutions can be made for a total crew or on an individual basis with minimum delay. Obviously, crew training is an indispensable factor in assuring the success of a mission.

Over 2,000 hours of formally scheduled training are required to develop the crews' operational and scientific abilities. This is equivalent to the classroom hours needed for a four-year college degree. These hours do not include the time of training-related activities for which a record cannot be readily established, such as study, physical exercise, informal briefings, and aircraft proficiency flying.

Many areas must be covered in training: the Saturn IB launch vehicle system and its functions; operation of the spacecraft, including the Orbital Workshop and all the experiments; medical training for those possible complications that can be diagnosed and treated in orbit; photography; moving outside the Skylab (EVA, Extravehicular Activity) and inside the Skylab (IVA, Intravehicular Activity) under the weightlessness of space; spacecraft fire training; inflight maintenance of Skylab systems; planetarium training for star fields, constellations, and specific celestial objects needed for navigation and for certain experiments; Skylab rescue operations; and modes of leaving the spacecraft under all preflight, inflight, and postflight ordinary and emergency situations on the launch complex, in flight, and in the water.

The astronauts received their training in four ways. First, they participated in numerous spacecraft and experiment tests conducted to insure faultless performance of all systems. This participation gave the crews valuable operating experience. Second, briefings on all Skylab systems were given to the astronauts in the form of lectures and demonstrations. Third, crew members participated in many of the reviews held during the development, manufacturing, and testing of Skylab components. On numerous occasions, astronauts suggested modifications of equipment and instruments which led to changes in the design of these components. The astronauts also took part in activities to develop operating procedures for instruments and systems. Fourth, the crews underwent systematic operational training, using training models and facilities, to learn how to perform all operational tasks, routine and emergency (Fig. 36).

Specially built simulators or trainers were used in many of these activities; by simulating major systems or components of Skylab, these training facilities enabled the crew members to practice operating procedures on earth (Fig. 37). A significant part of all crew training was done in these trainers.

The simulators included:

the Command Module Simulator which can simulate all the maneuvers of the actual Command Module;

the Skylab Simulator which was designed to provide systems and procedures training for Workshop functions;

the Command Module Procedures Simulator which was used primarily to develop skill in rendezvous and entry procedures;

the Dynamic Crew Procedures Simulator which supplemented the use of the Command Module Procedures Simulator and enabled the crews to practice launches and launch aborts.

Mock-ups, representing full-scale models of flight components, were constructed to establish or show dimensions and space requirements. These mock-ups, or trainers, were built from inexpensive materials; they usually did not contain operating subsystems. Among other purposes, mock-ups served the general familiarization of astronauts with Skylab components before the crews began simulator training. Skylab mock-ups included the Command Module Trainer, Multiple Docking Adapter Trainer, Airlock Module Trainer, and Orbital Workshop Trainer. (Fig. 38, 39, 40).

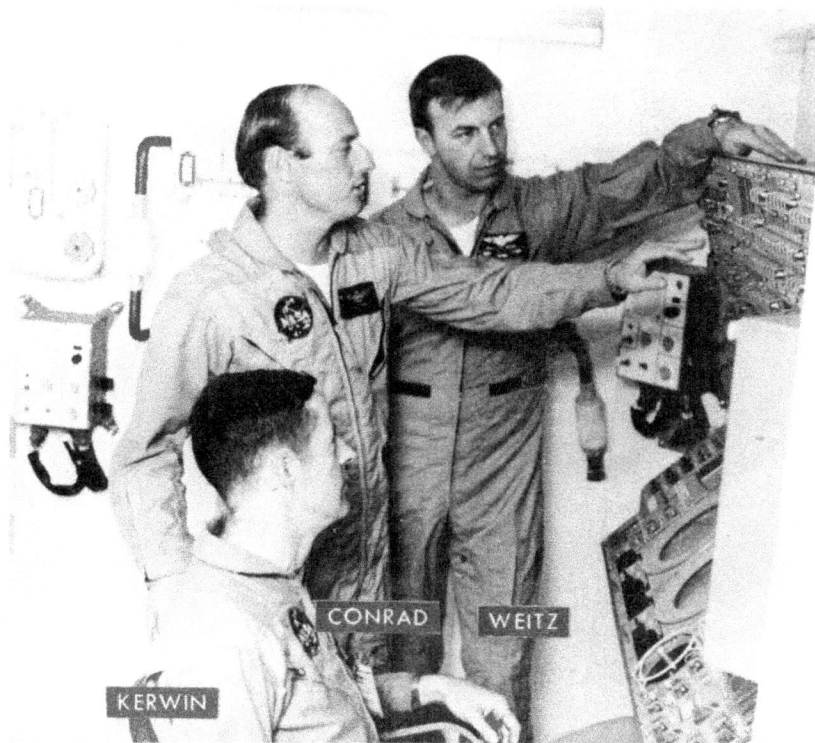

FIGURE 36.—Astronauts training in Skylab mockup.

Training for extravehicular activities and some intravehicular activities was performed to prepare the crew members for such tasks as leaving Skylab in the weightlessness of space in order to exchange film magazines in the cameras associated with ATM solar studies. Part of this training was carried out in airplanes of the KC–135 type which can generate a state of true weightlessness for periods up to about 30 seconds by flying through parabolic trajectories. Zero-gravity flights of this kind enabled crew members to practice such activities as eating and drinking, maneuvering, and tumble and spin recovery under weightless conditions.

Skylab crew members also trained in a simulated zero-gravity environment offered by the Neutral Buoyancy Space Simulator at the Marshall Space Flight Center (Fig. 41). This simulator is a water tank, 12 meters (40 ft) deep and 22.5 meters (75 ft) in diameter, containing wire mesh mock-ups of Skylab modules (Fig. 42). Full-size replicas of all four major elements of the Skylab cluster—the Workshop, the ATM, the MDA, and the Airlock—were submerged in the tank. Inside the tank, the astronauts wore pressurized space suits so weighted that they remained suspended in a relatively stable position, neither rising to the surface nor sinking to the bottom (Fig. 43, 44, and 45). Suspended in this "neutral buoyancy" condition,

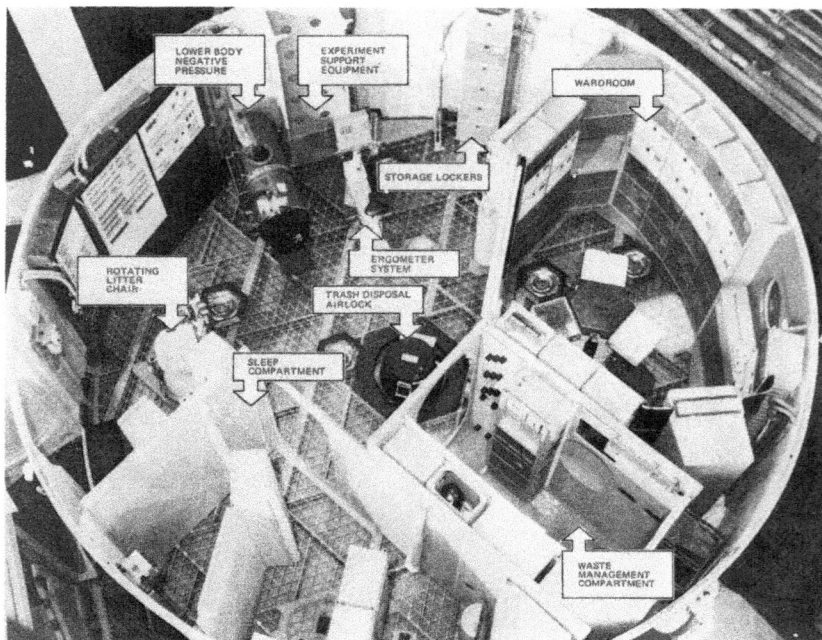

FIGURE 37.—Main floor Skylab Workshop mockup for training purposes.

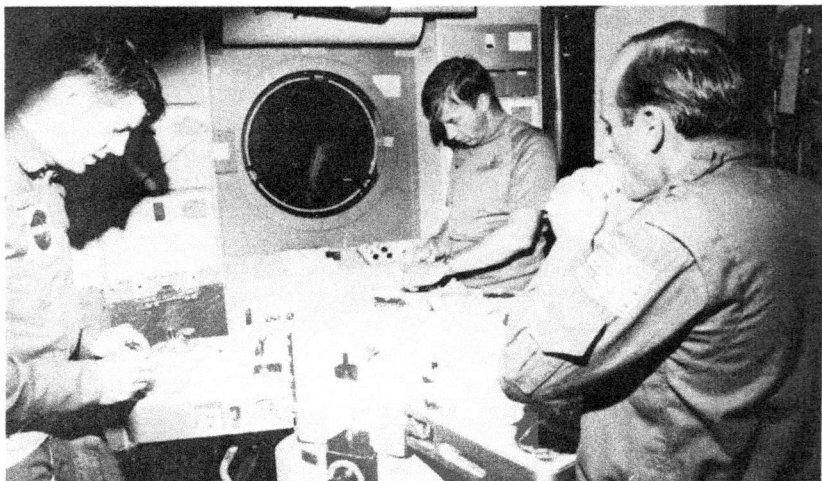

FIGURE 38.—Astronauts having meal at the food table in the wardroom trainer.

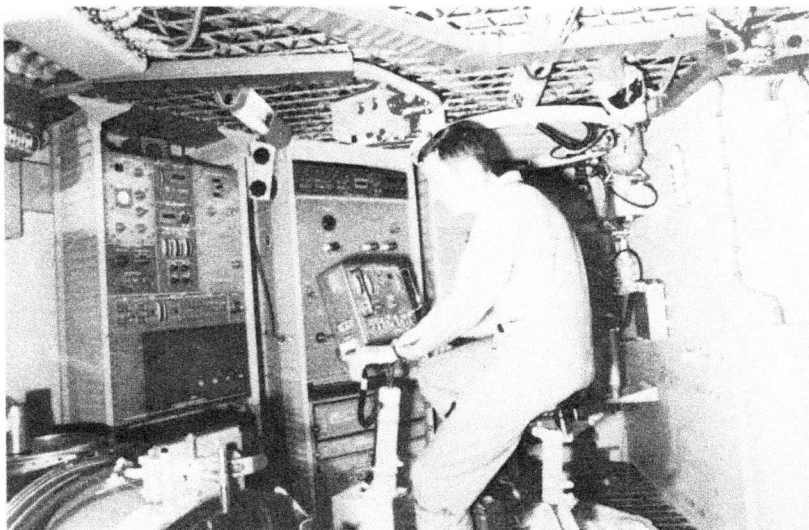

FIGURE 39.—Workshop Trainer with biomedical equipment.

the astronauts experienced some of the characteristic features of weightlessness, although neutral buoyancy provides, of course, only a limited simulation of true weightlessness in space. In this watery environment, the astronauts rehearsed many routine and some special tasks of the mission, particularly activities outside the Workshop.

The Skylab training program began with background training around November 1970. It provided general information and orientation on spacecraft systems and experiments, participation in spacecraft testing, reviews of flight plans and procedures, and training in solar physics.

Special training began in January 1972, for individual activities such as experiment operation, simulation of certain specific elements of the flight mission, and extravehicular and intravehicular activity training. Integrated crew training began with simulators and trainers around February, 1972, giving the astronauts and the flight controllers experience in flight and orbital operations. Finally, integrated mission team training began in November, 1972, to train each astronaut-controller team, and to give each astronaut experience in working together with the rest of the team. In this phase of training, the mission simulators were linked with the Mission Control Center, thus simulating the conditions of an actual mission.

7. SKYLAB CREW MEMBERS

Crews for Skylab were selected from the astronaut team, a group of men who are highly trained in many areas related to space flight (Fig 46). Additionally, Skylab crew members have received special training in all Skylab operations.

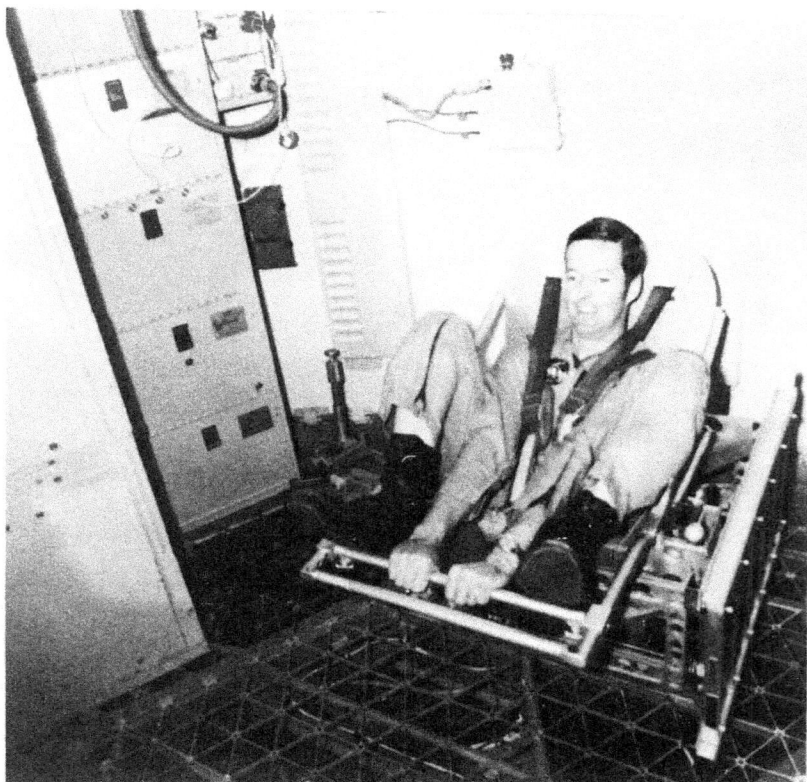

FIGURE 40.—Astronaut training on the body mass weighing device in the Workshop trainer.

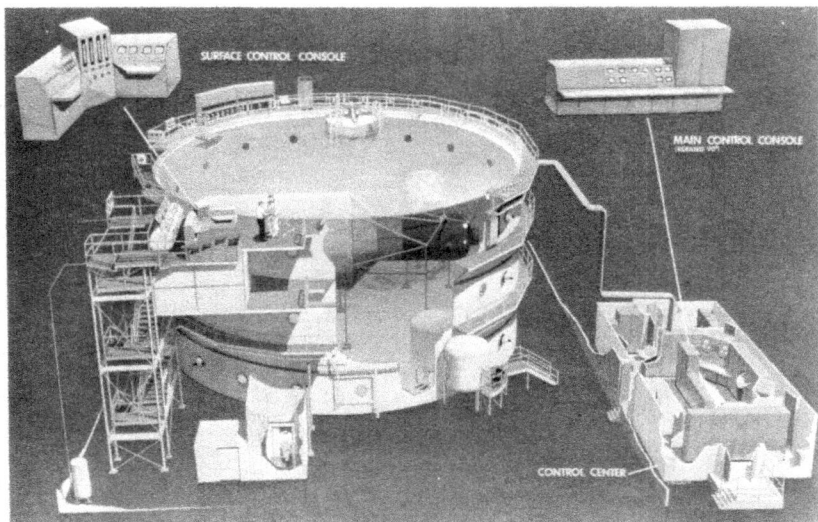

FIGURE 41.—View of the large water tank at the George C. Marshall Space Flight Center with a submerged Skylab mockup, used as a neutral buoyancy simulator.

FIGURE 42.—Full-size wire mesh mockup of Multiple Docking Adapter, used inside the water tank for neutral buoyancy training exercises.

FIGURE 43.—Two astronauts in space suits and a supporting diver during neutral buoyancy training exercises in water tank.

FIGURE 44.—Astronaut within Multiple Docking Adapter mockup during neutral buoyancy training exercises in water tank.

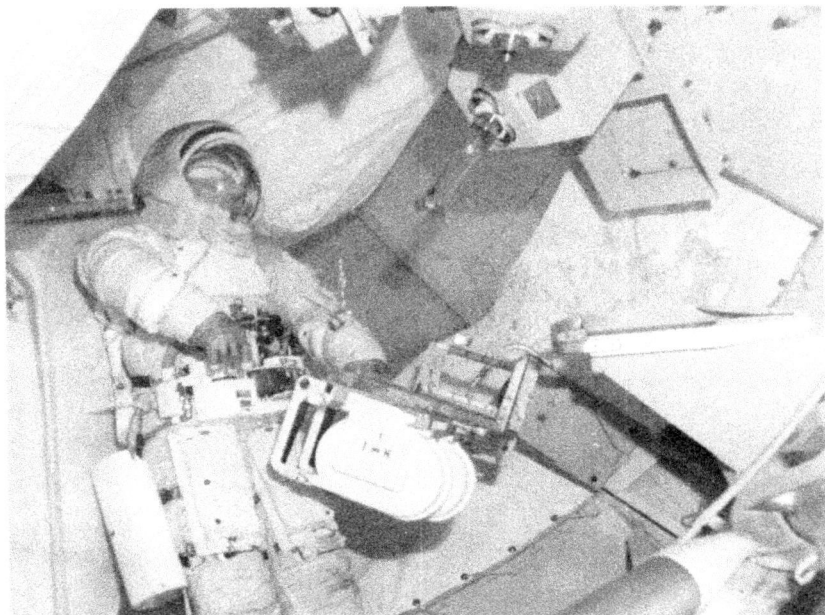

FIGURE 45.—Astronaut exchanging film camera during simulated Extravehicular Activity (EVA) under neutral buoyancy in water tank.

FIGURE 46.—Prime crewmen for the three manned Skylab missions.

There are three manned missions in the Skylab program. The first, beginning in May 1973, is a 28-day mission. Crew members for this mission are:

Charles Conrad, Jr., Commander (Fig. 47)
Dr. Joseph P. Kerwin, Science Pilot (Fig. 48)
Paul J. Weitz, Pilot (Fig. 49)

Backup crew:

Russell L. Schweickart, Commander (Fig. 50)
Dr. Story Musgrave, Science Pilot (Fig. 51)
Bruce McCandless II, Pilot (Fig. 52)

The second manned mission will begin in August 1973. This will be a 56-day mission. Crew members are:

Alan L. Bean, Commander (Fig. 53)
Dr. Owen K. Garriott, Science Pilot (Fig. 54)
Jack R. Lousma, Pilot (Fig. 55)

The backup crew, which also serves as backup crew for the third manned mission, consists of:

Vance D. Brand, Commander (Fig. 56)
Dr. William E. Lenoir, Science Pilot (Fig. 57)
Dr. Don L. Lind, Pilot (Fig. 58)

Crew members for the third manned mission, also a 56-day mission, are:

Gerald P. Carr, Commander (Fig. 59)
Dr. Edward G. Gibson, Science Pilot (Fig. 60)
William R. Pogue, Pilot (Fig. 61)

FIGURE 47.—Charles Conrad, Jr.

FIGURE 48.—Joseph P. Kerwin.

FIGURE 49.—Paul J. Weitz.

FIGURE 50.—Russell L. Schweickart.

FIGURE 51.——Story Musgrave.

FIGURE 52.——Bruce McCandless.

FIGURE 53.——Alan L. Bean.

FIGURE 54.——Owen K. Garriott.

FIGURE 55.—Jack R. Lousma.

FIGURE 56.—Vance D. Brand.

FIGURE 57.—William B. Lenoir.

FIGURE 58.—Don L. Lind.

FIGURE 59.—Gerald P. Carr.

FIGURE 60.—Edward G. Gibson.

FIGURE 61.—William R. Pogue.

First Skylab Mission

Charles (Pete) Conrad (Fig. 47) flew on Gemini 5 and 11 and on Apollo 12, the second manned lunar landing mission, for a total of 506 hours of space flight. He holds the rank of Captain in the U.S. Navy. He received a Bachelor of Science degree in Aeronautical Engineering from Princeton University in 1953; a Master of Arts degree from Princeton in 1966; an honorary Doctor of Laws degree from Lincoln-Wesleyan in 1970; and an honorary Doctor of Science degree from Kings College, Wilkes-Barre, Pennsylvania, in 1971. He was born on June 2, 1930, in Philadelphia, Pennsylvania.

This first Skylab mission will be the first flight in space for Dr. Joseph P. Kerwin (Fig. 48), a Commander in the Navy Medical Corps. He received a Bachelor of Arts degree in Philosophy from the College of the Holy Cross, Worcester, Massachusetts, in 1953, and a Doctor of Medicine degree from Northwestern University Medical School, Chicago, Illinois, in 1957. He served also as a naval aviator, earning his pilot's wings at Beeville, Texas, in 1962. He was born in Oak Park, Illinois, February 19, 1932.

The third crew member of Skylab's first manned mission, Paul J. Weitz (Fig. 49), is also a Commander in the U.S. Navy. As a naval aviator, he received five awards of the Air Medal and the Navy Commendation Medal for combat flights in the Vietnam area. He received a Bachelor of Science degree in Aeronautical Engineering from Pennsylvania State University in 1954 and a Master's degree in Aeronautical Engineering from the U.S. Naval Postgraduate School in Monterey, California, in 1964. He was born in Erie, Pennsylvania, on July 25, 1932. He has not flown in space before.

Backup Crew, First Manned Mission

Russell L. (Rusty) Schweickart (Fig. 50) served as the lunar module pilot on Apollo 9, March 3–13, 1969. This was the third manned flight in the Apollo series, the second to be launched by a Saturn V, and the first manned flight of the lunar module. Although he served as a pilot in the U.S. Air Force from 1956 to 1960 and was recalled to active duty for a year in 1961, he is not now a member of the military forces. Before joining NASA in 1963, he was a research scientist at the Experimental Astronomy Laboratory at the Massachusetts Institute of Technology (MIT). He received a Bachelor of Science degree in Aeronautical Engineering in 1956 and a Master of Science degree in Aeronautics and Astronautics in 1963 from MIT. He was born October 25, 1935, in Neptune, New Jersey.

Dr. Story Musgrave (Fig. 51) has earned five college degrees; a Bachelor of Science degree in Statistics from Syracuse University in 1958; a Master's degree in Business Administration in Operations Analysis from the University of California, Los Angles, in 1959; a Bachelor of Arts degree in Chemistry from Marietta College in 1960; an M.D. from Columbia University in 1964; and a Master of Science Degree in Biophysics from the University of Kentucky in 1966. He has flown over 30 types of aircraft and holds instructor, instrument instructor, and airline transport ratings. He, like Russell Schweickart, is a "civilian" astronaut, not a member of the military forces. He has not yet flown in space. He was born August 19, 1935, in Boston, Massachusetts.

Bruce McCandless II (Fig. 52) is a Lieutenant Commander in the U.S. Navy. He has a Bachelor of Science degree in Naval Sciences from the U.S. Naval Academy, received in 1958; a Master of Science degree in Electrical Engineering from Stanford University, received in 1965. He was designated a Naval Aviator in March of 1960. He was born June 8, 1937 in Boston, Massachusetts. He has not flown in space as yet.

Second Skylab Mission

Alan L. Bean (Fig. 53) is a Captain and a pilot in the Navy. He was the lunar module pilot of Apollo 12, November 14–24, 1969, a mission that lasted 244 hours and 36 minutes. During this mission, Captain Bean spent 7 hours and 45 minutes EVA on the lunar surface. He holds a Bachelor of Science degree in Aeronautical Engineering from the University of Texas, received in 1955. He was born in Wheeler, Texas, on March 15, 1932.

Dr. Owen K. Garriott (Fig. 54), Science Pilot on the second manned mission, holds a Doctorate in Electrical Engineering from Stanford University, received in 1960. He also has a Bachelor of Science degree in Electrical Engineering from the University of Oklahoma, received in 1953, and a Master of Science degree in Electrical Engineering from Stanford, received in 1957. He has not yet flown in space. Dr. Garriott was born November 22, 1930, in Enid, Oklahoma.

Jack Robert Lousma (Fig. 55) is a pilot holding the rank of Major in the Marine Corps. He received a Bachelor of Science degree from the University of Michigan in 1959 and the degree of Aeronautical Engineer from the U.S. Naval Postgraduate School in 1965. He was born February 29, 1936, in Grand Rapids, Michigan. He has not flown in space before.

Backup Crew, Second and Third Manned Missions

Vance D. Brand (Fig. 56), Commander for the backup crew for these missions, was a commissioned officer and naval aviator with the Marine Corps from 1953 to 1957. When he joined NASA, he was an experimental test pilot and leader of a Lockheed Aircraft Corporation flight test advisory group. He holds a Bachelor of Science degree in Business from the University of Colorado, received in 1953; a Bachelor of Science degree in Aeronautical Engineering from the University of Colorado in 1960; and a Master's degree in Business Administration from the University of California at Los Angeles in 1964. Mr. Brand was born in Longmont, Colorado, on May 9, 1931. Should he be called upon to fly on Skylab, he will have his first mission in space.

Dr. William E. Lenoir (Fig. 57), Science Pilot, is a graduate of the Massachusetts Institute of Technology, where he received a Bachelor of Science degree in 1961, a Master of Science degree in 1962, and a Doctorate in 1965. He has served as instructor at MIT and in 1965 was named Assistant Professor of Electrical Engineering. He is acting as an investigator in several satellite experiments and has continued his research in the areas of space engineering and physics while serving as an astronaut. He has not yet flown in space. Dr. Lenoir was born on March 14, 1939, in Miami, Florida.

Dr. Don L. Lind (Fig. 58), Pilot, before his selection as an astronaut in 1966, worked as a space physicist at the NASA Goddard Space Flight Center. He is a former naval aviator, earning his wings in 1955. Dr. Lind received a Bachelor of Science degree with high honors in Physics from the University of Utah in 1953 and a Doctor of Philosophy degree in High Energy Nuclear Physics in 1964 from the University of California at Berkeley. He was born May 18, 1930, in Midvale, Utah. He has not yet flown in space.

Third Manned Mission

Gerald P. Carr (Fig. 59) is a Marine Corps Lieutenant Colonel and a Marine pilot. He received a Bachelor of Science degree in Mechanical Engineering from the University of Southern California in 1954, a Bachelor of Science degree in Aeronautical Engineering in 1961 from the U.S. Naval Postgraduate School, and a Master of Science degree from Princeton in 1962. He was born in Denver, Colorado, on August 22, 1932. This will be his first flight in space.

The Science Pilot for the third manned mission is Dr. Edward G. Gibson (Fig. 60), who received his Doctorate in Engineering degree with a minor in Physics from the California Institute of Technology in June 1964. He received a Bachelor of Science degree in Engineering from the University of Rochester, New York, in June 1959, and a Master of Science degree in Engineering (jet propulsion option) from the California Institute of Technology in June 1960. After his selection as an astronaut in 1965, he completed his Air Force flight training in 1966. Dr. Gibson was born November 8, 1936, in Buffalo, New York. He has not flown in space before.

William R. Pogue (Fig. 61) is a Lieutenant Colonel and pilot in the U.S. Air Force. He received a Bachelor of Science degree in Education from Oklahoma Baptist University in 1951 and a Master of Science degree in Mathematics from Oklahoma State University in 1960. He flew 43 combat missions with the Fifth Air Force in 1953 and 1954 during the Korean conflict. From 1955 to 1957, he was a member of the U.S. Air Force Thunderbirds precision flying team. Colonel Pogue was born January 23, 1930, in Okemah, Oklahoma. This mission will be his first space flight.

8. LAUNCH PREPARATIONS

Final assembly of the complete Skylab cluster took place in the Kennedy Space Center early in 1973. Airlock Module and Multiple Docking Adapter, after being joined together at the McDonnell Douglas Corporation in St. Louis, were flown to KSC in the new Commercial Guppy airplane designed specifically to fly large cargoes. The plane, under contract to NASA, is operated by Aero Spacelines, Inc., Santa Barbara, California. The Super Guppy (Fig. 62), somewhat smaller than the new Guppy, transported the Command and Service Module from Downey, California, to KSC on July 18, 1972, and the Apollo Telescope Mount from JSC, where thermo-vacuum testing was performed, to KSC on September 22, 1972. The Instrument Unit was shipped by Super Guppy from the George C. Marshall Center to KSC

500-721 O - 73 - 5

FIGURE 62.—Air transport plane Super Guppy, transporting large Skylab components from manufacturing and test sites to Kennedy Space Center, Florida.

on October 26, 1972. The four solar arrays for the ATM arrived at KSC in two separate trips in mid-December, 1972.

The Orbital Workshop with its solar power arrays, too big for air transport, was loaded on a specially-equipped ocean-going vessel provided by the U.S. Navy's Military Sealift Command, the USNS *Point Barrow* (Fig. 63) and shipped from Seal Beach, California, to Port Canaveral, Florida, through

FIGURE 63.—USNS Point Barrow transporting large Skylab components from manufacturing and test sites to the launch site at Cape Kennedy, Florida.

the Panama Canal. The Payload Shroud was part of the same shipment. The voyage took approximately 14 days.

Assembly and successive testing procedures of the Skylab cluster are described in Chapter IV–1. The Skylab spacecraft was "stacked" on the Saturn V launch vehicle in the Vertical Assembly Building at KSC (Fig. 64) late in January; meanwhile, the CSM was mounted on the Saturn IB launch vehicle. The components of the Saturn V and Saturn IB launch vehicles had arrived at KSC by covered barge from the Michoud Plant in New Orleans on July 26, 1972, and on August 22, 1972. Integrated space vehicle testing of Skylab, including simulated flight tests of all systems, was performed during the month of February (Fig. 65). Early in April, the huge Saturn V launch vehicle with its payload was transported by crawler from the Vertical Assembly Building to the launch pad (Fig. 66). The Saturn IB arrived on its launch pad in March. The crawler transporter is a very large tracked vehicle (Fig. 67). It has a flat top surface 40 meters (131 ft) long and 34.8 meters (114 ft) wide on which it carries Saturn launch vehicles to the pads. Final tests of all components, and of the complete systems, extended over a period of two-and-one-half months in the Vertical Assembly Building. Particular care had to be taken during the last phase on the pad that the various instruments, particularly the solar telescopes in the ATM canister, did not suffer from environmental contamination.

Countdown for launch will begin about one week before lift-off time. The major steps of the countdown, such as fueling, battery charging, pressurization of the Workshop, and final checks, are listed in Table 4. The complete countdown list as used for the actual launching is a book with about 200 pages and 1500 individual line items (sequential operations).

FIGURE 64.—Vertical Assembly Building at Kennedy Space Center, Florida.

Figure 65.—Inside Vertical Assembly Building with the Orbital Workshop being put atop the Saturn V vehicle.

FIGURE 66.—Launch Pad 39A used for the launching of Saturn V vehicles.

FIGURE 67.—Crawler vehicle used for transportation of Saturn IB and Saturn V
launch vehicles from the Vertical Assembly Building to the launch pads.

TABLE 4.—*Major Steps of Countdown for Launch* (*Times before Liftoff*)

Action	Days	Hrs.	Min.	Sec.
Arrival on pad—				
Saturn IB on pad 39B......................	71	0	0	0
Saturn V on pad 39A......................	30	0	0	0
Storage bottles for gaseous oxygen and nitrogen filled...................................	18	0	0	0
Countdown demonstration test completed.......	12	0	0	0
Move mobile service structure from pad 39A to pad 39B for first manned launch.............	6	0	0	0
Installation of ordnance (explosive charges) completed...............................	3	2	30	0
Installation and activation of stage batteries.....	2	10	0	0
Validation of communication link for launch support completed.........................		20	0	0
Clearing of launch pad.........................		6	30	0
Launch vehicle propellant loading start................		5	30	0
Range safety checks............................		4	0	0
Thruster attitude control system (TACS) cover jettison..		2	30	0
Built-in hold..		2	0	0
Launch vehicle power transfer test............................			39	0
Spacecraft (Skylab) switched to internal power....................			8	0
Spacecraft (Skylab) final status checks.........................			3	7
Automatic therminal sequence start (firing command)...			3	7
Launch vehicle transfer to internal power.............................				50
Verify launch sequence...				30
Retraction of Saturn first stage (SIC) forward swing arm..				16. 2
Final checkout of systems (verify Saturn first stage [SIC] engine thrust)...				1. 9
Launch commit...				T−0
Liftoff (first motion)...				T+0. 3

CHAPTER IV

Skylab Design and Operation

Historically, the design of Skylab and the concept of its functions have evolved through several phases of increasing complexity. From the beginning, however, Skylab was to be a laboratory in orbit where men could live and work for extended periods. Man was to contribute to the accomplishments of Skylab in four major ways: as a scientific observer, as an experimenter, as an operator, and as the object of biomedical studies. The Skylab Project will offer its crew members the opportunity of intimate involvement in these four areas of activity.

It will form the basis from which many of the future space systems for science, technology, Earth observations, applications, and exploration will evolve.

The following sections describe details of the design and the functions of the major Skylab system components, and of the subsystems needed for Skylab operation.

1. SKYLAB COMPONENTS

The Skylab cluster, first U.S. space station, will be the largest manned spacecraft ever placed in Earth orbit. With the Command and Service Module docked to it, the Skylab cluster will be approximately 35 meters (117 ft.) long; it will have a mass of 90,606 kilograms (199,750 lbs.) and contain a habitable volume of about 354 cubic meters (12,700 cubic ft.) (Fig. 69). Major Skylab cluster elements include the following:

The Orbital Workshop (OWS) houses the crew quarters, most of the stored expendables, and a large experiment area. It supports the large solar arrays, and it contains the cold gas tanks and thrusters for secondary attitude control.

The Airlock Module (AM) has an airlock for extravehicular activities, the main systems for communication and data transmittal, environmental and thermal control systems, and the electric power control system.

The Multiple Docking Adapter (MDA) provides the docking ports for the Command and Service Module (CSM); it houses the control console for the Apollo Telescope Mount (ATM), controls and sensors for Earth resources viewing, and a number of other experimental facilities.

The Apollo Telescope Mount carries the solar telescopes, the control moment gyros (CMG) for primary attitude control, and four solar array wings.

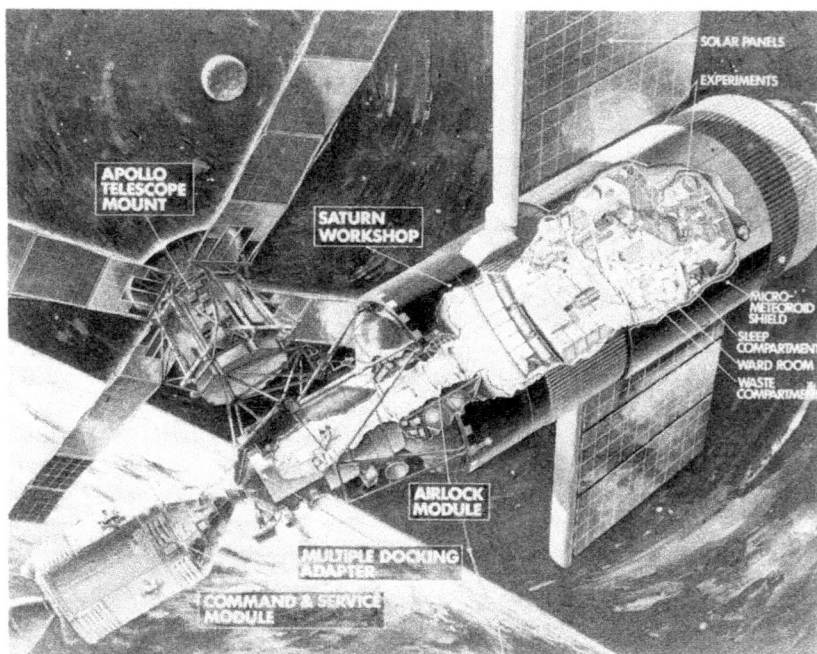

FIGURE 69.—Skylab cluster in orbit, showing major components and interior equipment.

The modified Command and Service Module functions as the manned logistics vehicle for the missions and also provides certain communication functions to the Workshop.

Each major Skylab cluster element is described in detail in the following section. Fig. 70 depicts the individual elements of the cluster.

a. Orbital Workshop (OWS)

From the outside, the Orbital Workshop looks in size and shape like the S–IVB stage which served as the second propulsion stage of the Saturn IB launch vehicle and as the third propulsion stage of the Saturn V-Apollo launch vehicle (Fig. 71). Attached to the outside wall are two large, wing-like solar arrays. The interior of this stage, which originally consisted mainly of two tanks for liquid hydrogen and liquid oxygen, underwent a thorough conversion; the larger hydrogen tank became a living and working facility (the Workshop) for three astronauts, and the smaller oxygen tank became a container for waste products accumulating during the mission (Fig. 72). The Workshop measures 6.7 m (22 ft) in diameter, and 14.6 m (48 ft) in length. Its habitable volume is 275 m³ (9550 ft³), its weight 35,380 kg (78,000 lbs). The solar arrays extend 9 m (30 ft) on opposite sides for a total tip-to-tip wing span of about 27 m (90 ft).

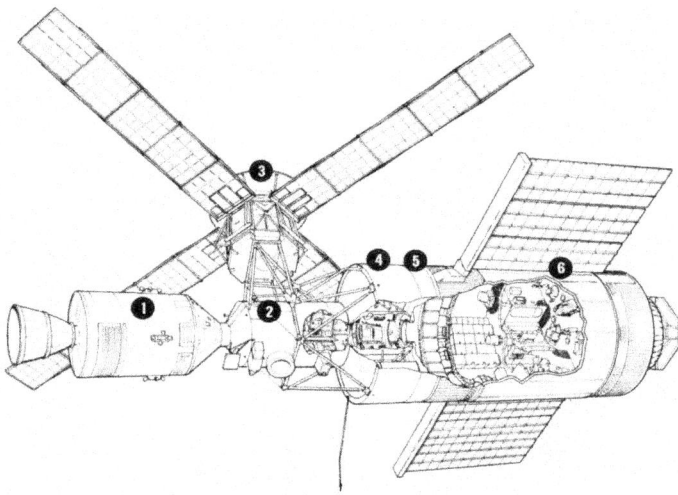

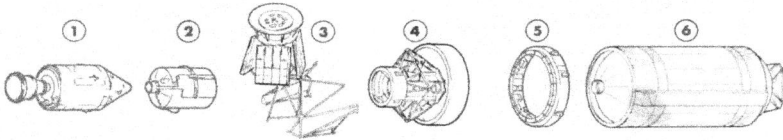

FIGURE 70.—Major parts of Skylab cluster.

To protect the Workshop against penetration by meteoroids, a thin metal shield envelops the Workshop wall at a distance of 0.15 meter (6 inches) from the outer surface. Meteoroid particles hitting this shield will suffer an energy reduction, and they will be broken up into a shower of numerous smaller particles, none of which will be able to penetrate the Workshop wall. During launch, the meteoroid shield will be packed tighty against the wall; after orbit insertion, the shield is swung out into position by torsion bars.

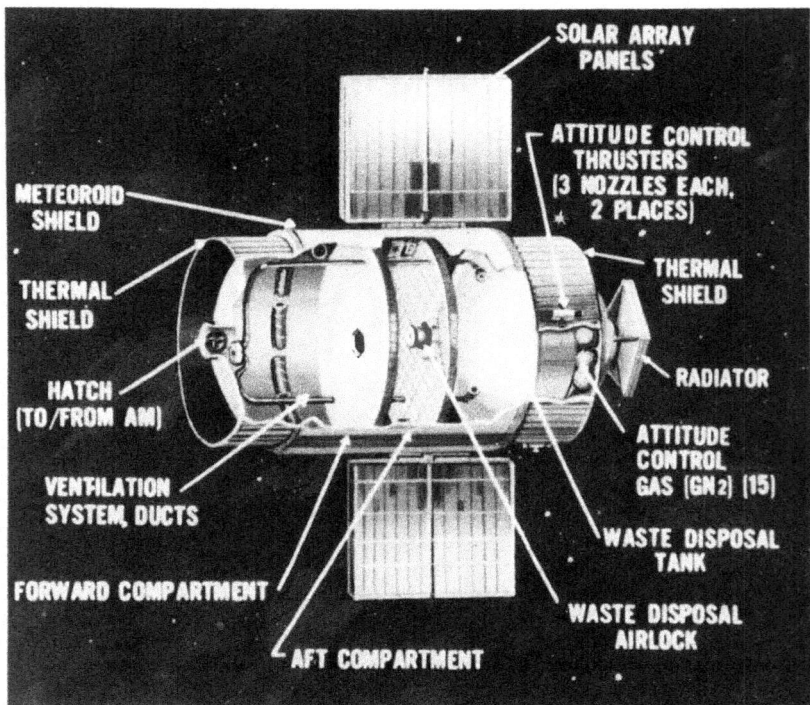

FIGURE 71.—Basic elements of Orbital Workshop (OWS).

Serving as living quarters and as a laboratory for the astronauts, the Workshop consists of two compartments, separated by a perforated wall or "floor." The rearward compartment contains the wardroom for food preparation and eating, the sleep section, the waste management section (toilet), and the experiment work area. Biomedical experiments will be performed in this work area (Fig. 73). The forward compartment is devoted primarily to experiments requiring relatively large volumes or designed to utilize one of the two scientific airlocks for external viewing or exposure (Fig. 74).

Storage facilities, containers for food, water, and clothing, and a number of subsystems occupy part of the room available in the two compartments.

Circulation of the air will be achieved by fans and by ducts between the Workshop wall and a partial inner wall which will permit the air to flow in one direction between the walls and in the other direction through the room.

A number of handles, grips, and foot restraints, mounted at appropriate places throughout the two compartments, will enable the crew members to move through the Workshop and to station themselves at desired locations while floating weightlessly (Fig. 75).

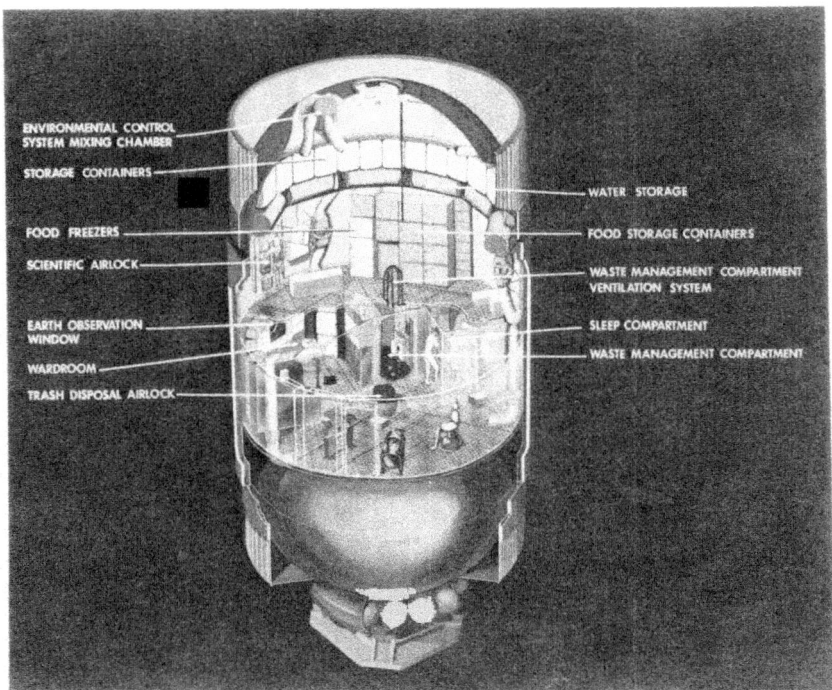

ENVIRONMENTAL CONTROL
SYSTEM MIXING CHAMBER

STORAGE CONTAINERS

FOOD FREEZERS

SCIENTIFIC AIRLOCK

EARTH OBSERVATION
WINDOW

WARDROOM

TRASH DISPOSAL AIRLOCK

WATER STORAGE

FOOD STORAGE CONTAINERS

WASTE MANAGEMENT COMPARTMENT
VENTILATION SYSTEM

SLEEP COMPARTMENT

WASTE MANAGEMENT COMPARTMENT

FIGURE 72.—Arrangement and major components of interior installations of Orbital
Workshop, showing the two floors of the living area.

A window in the wardroom will allow the astronauts to look out in a direction away from the Sun. Depending on Skylab's location in orbit, they will be able to look down at the Earth or out into space.

Two scientific airlocks are provided in the Workshop wall of the forward compartment, one directed toward the Sun, the other facing in the opposite direction (solar and antisolar airlock) (Fig. 76). Each of the scientific airlocks will permit deployment of sensors and instruments such as Experiments No. S019, T027, and S073 (see Chapter V). An articulated mirror system and a universal extension mechanism can be used with the scientific airlocks for certain experiments (Figs. 77, 78).

A considerable array of tools, supplies, and support equipment will be available to the astronauts on Skylab, including repair kits, films, handtools, tapes, clamps, fasteners, scissors, thermometers, photographic lights, and still and movie cameras.

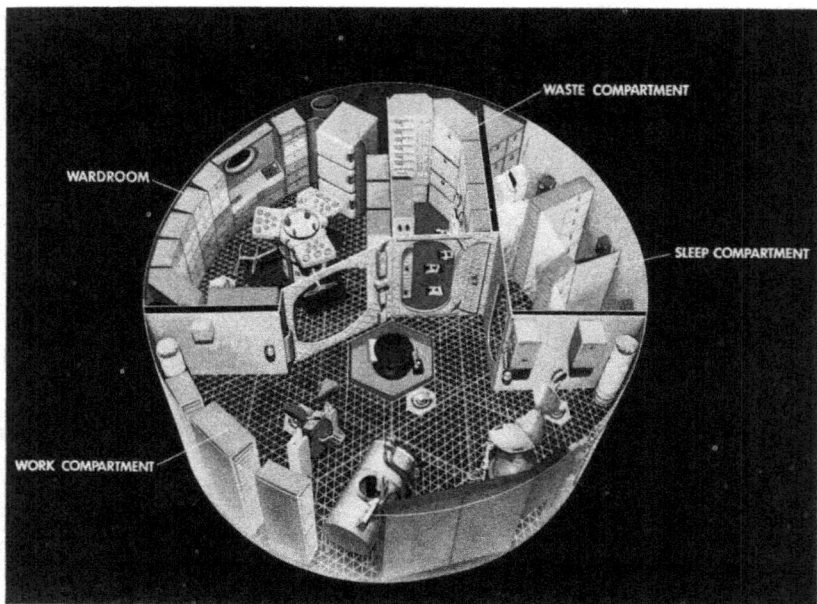

FIGURE 73.—First (lower) floor of Orbital Workshop, showing crew quarters and medical experiment work area.

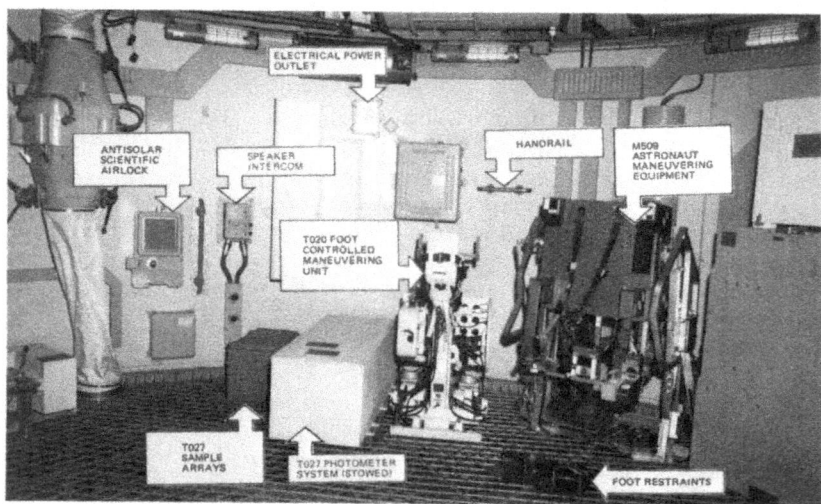

FIGURE 74.—Second (upper) floor of Orbital Workshop, showing equipment for space technology experiments.

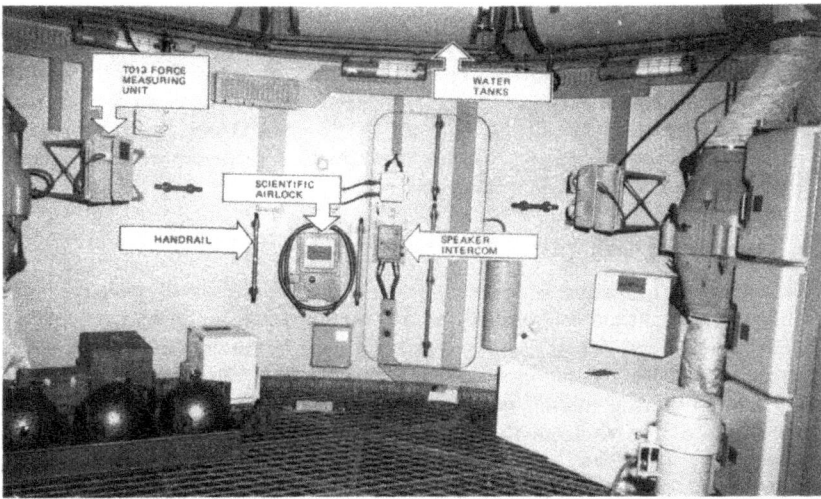

FIGURE 75.—Second (upper) floor of Orbital Workshop, showing scientific airlock, hand rails, and other installations.

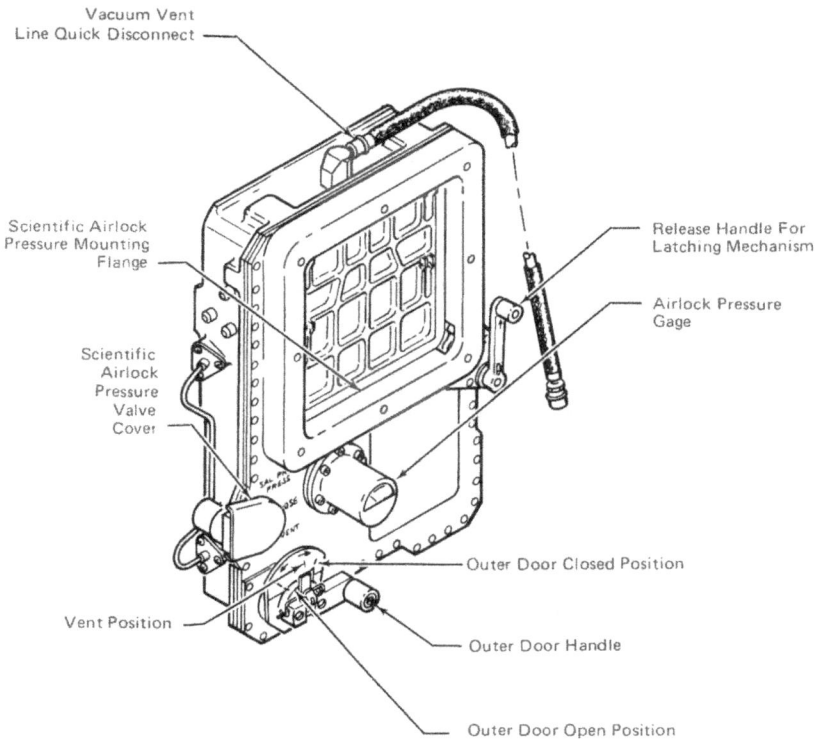

Vacuum Vent
Line Quick Disconnect

Scientific Airlock
Pressure Mounting
Flange

Scientific
Airlock
Pressure
Valve
Cover

Vent Position

Release Handle For
Latching Mechanism

Airlock Pressure
Gage

Outer Door Closed Position

Outer Door Handle

Outer Door Open Position

FIGURE 76.—Inside door of scientific airlock.

Skylab's atmosphere will be a mixture of 25,000 Nm⁻² (Newtons per square meter) (0.25 atm. or 3.7 psi) partial pressure oxygen and 9,000 Nm⁻² (0.09 atm. or 1.3 psi) partial pressure nitrogen. Relative humidity will be controlled to about 26% at 29° C (85° F); room temperature can vary between 13° C and 32° C (55° F and 90° F) (See Chap. IV–2–b).

The workshop is connected with the rest of the Skylab cluster through the Instrument Unit, as illustrated in Fig. 79.

b. Instrument Unit (IU)

Control of the Saturn V launch vehicle during launch and powered flight will be accomplished by guidance and control systems located in the Instrument Unit (Figs. 79, 80). This function will be maintained by the IU throughout Skylab orbit insertion and deployment. Equipment in the IU will first guide the launch vehicle from the moment of liftoff through the separation of Skylab from the second stage of the Saturn V booster. After separation, the IU will provide commands to various Skylab systems which in turn will rotate the Skylab by 180°, turn on refrigeration systems, jettison the payload shroud, roll the Skylab until the Apollo Telescope Mount points toward the Sun, deploy the meteoroid shield that envelops the Workshop, and pressurize all the compartments with oxygen (Skylab will be filled only with nitrogen during launch). The solar arrays on ATM and on OWS will be deployed upon command from the IU or by command from the ground. All these functions will be completed about 7.5 hours after orbit insertion. The batteries energizing the IU will be depleted soon after this time. From then on, the Instrument Unit will be passive.

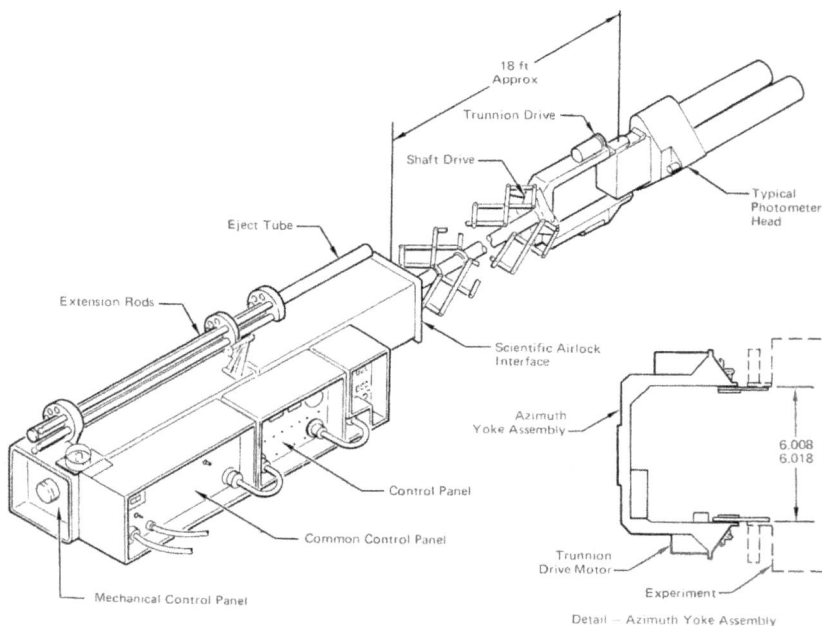

FIGURE 77.—Expanded view of scientific airlock, showing photometer deployment.

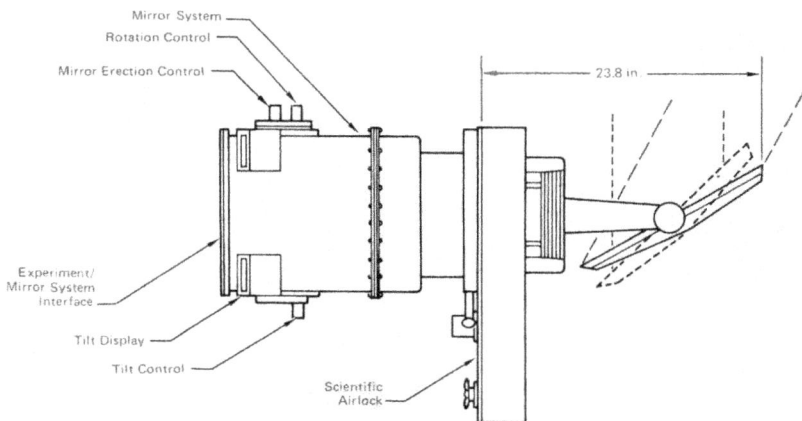

FIGURE 78.—Movable mirror outside the airlock, as needed for certain scientific observations.

c. Airlock Module (AM)

The Airlock Module, as a connecting link between OWS and Multiple Docking Adapter, serves a threefold purpose: as a major structural element of the Skylab cluster; as a module containing the port through which an astronaut can leave the interior of Skylab in order to perform extravehicular activities (EVA); and as the electrical, environmental, and communications control center for Skylab. In addition, many of the high pressure containers for oxygen and nitrogen which provide Skylab's atmosphere are mounted on the trusses between the inner and outer walls of the AM.

The Airlock Module consists of two concentric cylinders (Fig. 8). Matching the OWS in diameter, the outer cylinder or Fixed Airlock Shroud carries the Payload Shroud during the launch, and it serves as mounting base for the structure that supports the Apollo Telescope Mount. The inner cylinder, or tunnel, represents the airlock (Figs. 81, 82). It forms the passageway through which crew members can move between the Workshop and the Multiple Docking Adapter. Hatches at both ends of the tunnel can be closed for depressurization (Figs. 83, 84), and a third hatch in the side wall can be opened for the egress of a crew member (Fig. 85). After return of the crew member, the egress hatch is closed, the tunnel is pressurized, and the forward and rear hatches are reopened.

The Airlock Module also contains the automatic Skylab malfunction alarm system, and the manual controls for Skylab pressurization and air purification, and for electric power and communications.

Many of the supplies, and most of the control systems for Skylab are located in the Airlock Module; this module could well be called the "utility center" of the Skylab cluster.

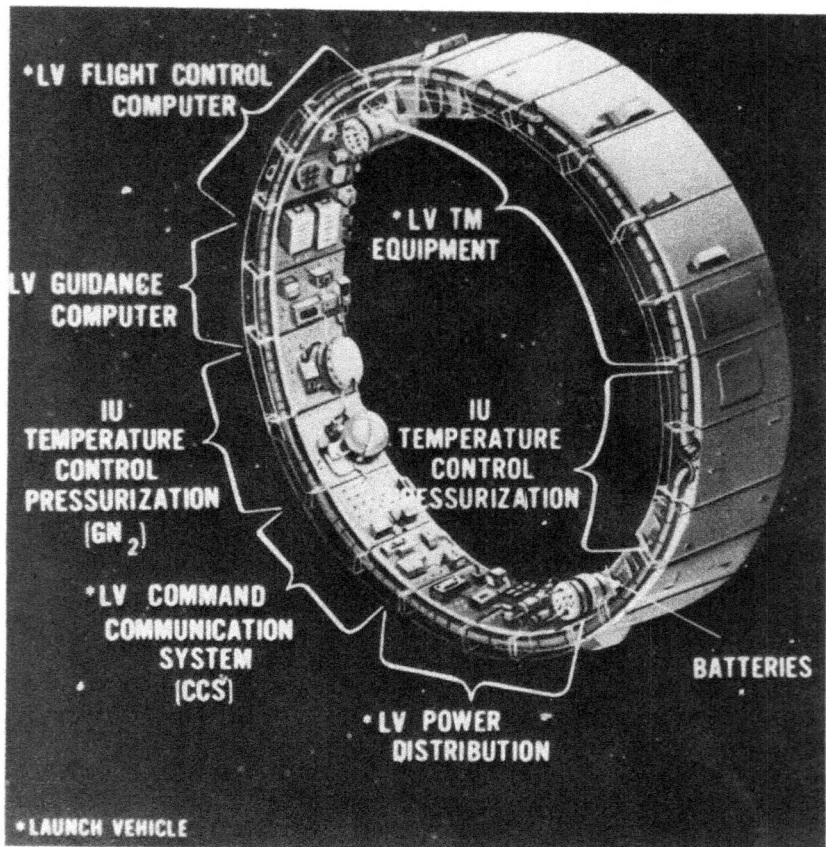

FIGURE 79.—Instrument Unit, showing guidance, control, and power systems equipment.

d. Multiple Docking Adapter (MDA)

Like other major components of Skylab, the Multiple Docking Adapter will serve several purposes. It is equipped with docking ports for the manned Command and Service Modules that will carry astronauts to and from Skylab. It houses the control units for ATM, for the Earth Resources Experiment Package (EREP), and for the M512 (zero-gravity materials processing facility); and it is used for storage of films, experiment components, and electric and television equipment (Fig. 86).

Two docking ports are provided on the MDA. The primary port, located at the forward end for axial docking of a CSM, will be used under all normal conditions. The side port will be used in contingency cases only.

EREP equipment, consisting of sensors and cameras for Earth viewing and of control and display units to be operated by the astronauts, is mounted in the MDA. These Earth-observing instruments will be looking in the antisolar direction (Figs. 87, 88).

FIGURE 80.—Instrument Unit during assembly.

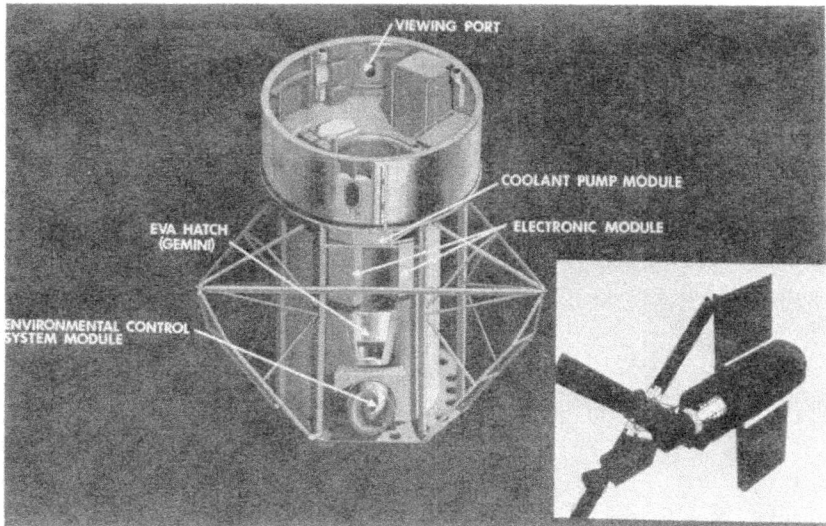

FIGURE 81.—Airlock Module with hatch for extra-vehicular activities.

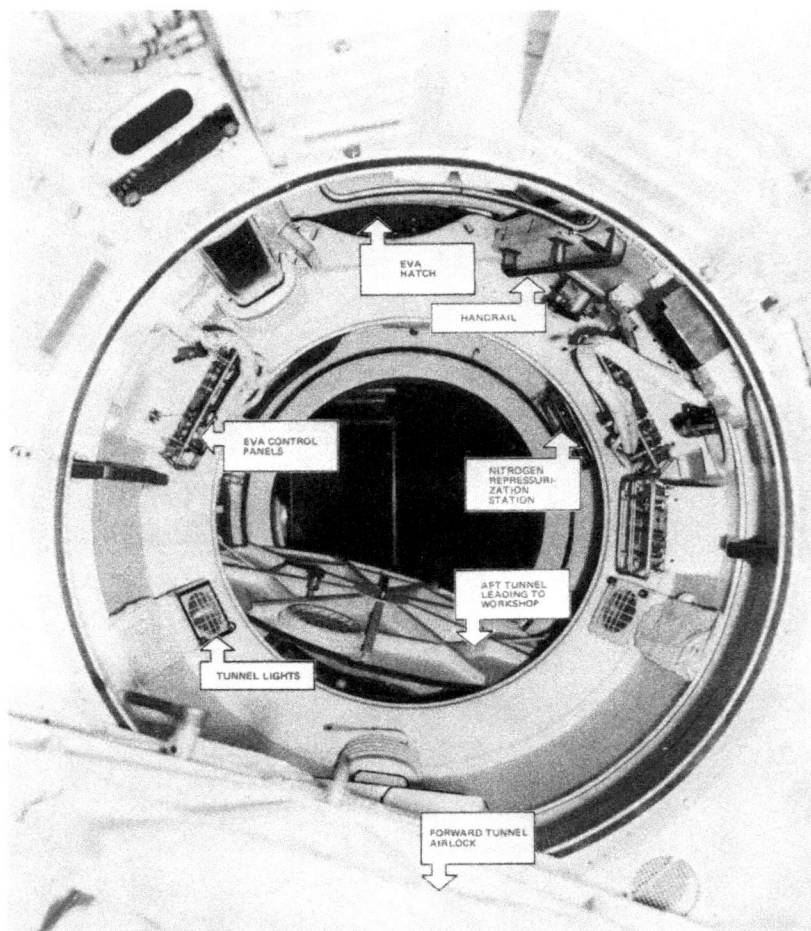

FIGURE 82.—Inside view of Airlock Module with open door leading to Workshop.

Astronaut operation of the Apollo Telescope Mount and the Skylab attitude control will be performed at the ATM Control and Display Console in the MDA. This console contains several TV screens and other visual indicators which will enable the astronauts to follow ATM activities closely and to actively participate in target selection, telescope orientation, experimental procedures, and interpretation of instrumental and visual observations.

The television system for Skylab is housed in the MDA. Several remote stations are located in the other Skylab modules. A video tape recorder is included in this system.

Crew members will spend a significant portion of their orbital time in the Multiple Docking Adapter, at the ATM and EREP consoles and the M512 zero-gravity materials processing and manufacturing facility.

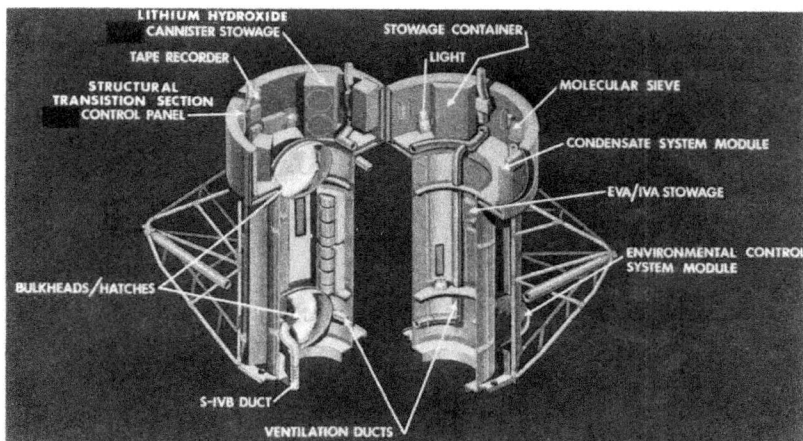

FIGURE 83.—Airlock Module, showing major components and installations.

e. Apollo Telescope Mount (ATM)

During launch and ascent, the Apollo Telescope Mount unit is positioned axially with the rest of Skylab and the launch vehicle (Fig. 11). After insertion into orbit and jettisoning of the shroud (Fig. 89), the ATM support structure will rotate the ATM by 90° from its axial position into a radial position (Fig. 90). This operation will clear the primary docking hatch, and it will enable the ATM telescopes and the ATM solar arrays to face the Sun. The ATM solar panels will be deployed after this 90° rotation.

ATM consists of two major parts, the outer structure or rack and the inner part or canister with the solar telescopes (Fig. 91). The rack, an octagonal truss structure of 3.3 m (11 ft) diameter and 3.6 m (12 ft) length, connects ATM with the rest of Skylab. It carries the four solar-electric power arrays and electric batteries. The rack also contains electrical and mechanical (gyro) components of the Skylab primary attitude control system and the ATM communications system.

Attitude control of Skylab must be achieved with a system which operates for a long period of time, while producing as little environmental contamination as possible because of the sensitivity of the solar, stellar, and Earth-looking telescopes to condensed vapors and particulate matter in the vicinity of the spacecraft. For these reasons, control moment gyroscopes (CMG) were chosen to produce the major portion of required control torques. The three control moment gyros are mounted on the rack (Fig. 92). The rack and the rest of Skylab, being rigidly connected to each other, will act as one unit as far as primary attitude control is concerned.

Inside the rack, the cylindrical canister is mounted in such a way that it can be rotated around its cylindrical axis upon a manual signal. This roll ring mount (Fig. 93) provides the ±120° roll capability which is required for polarization studies of the solar radiations, and also for the exchange of some of the film cassettes. The astronaut, in order to retrieve the cassettes,

Ventilation Ducts
Electrical Outlet
Molecular Sieve
Flight Spares Stowage Container
Film Tree Support
Extravehicular Activity Hatch
Light and Handrail Assembly
Duct To Orbital Workshop

Window
O$_2$/N$_2$ Control Panel
Permanent Stowage Container
Power and O$_2$ Connector
Tape Recorder Module
TV Input Station
Instrument Panels

Control Panel
Electrical Outlets
Panel Lights
Lock Compartment Hatch
Tunnel Lights
Recharge Provision for Nitrogen Tanks

Window
Teleprinter Paper Stowage Container
Electrical Outlet
Apollo Telescope Mount Film Magazine Temporary Storage
Lock Compartment Hatch

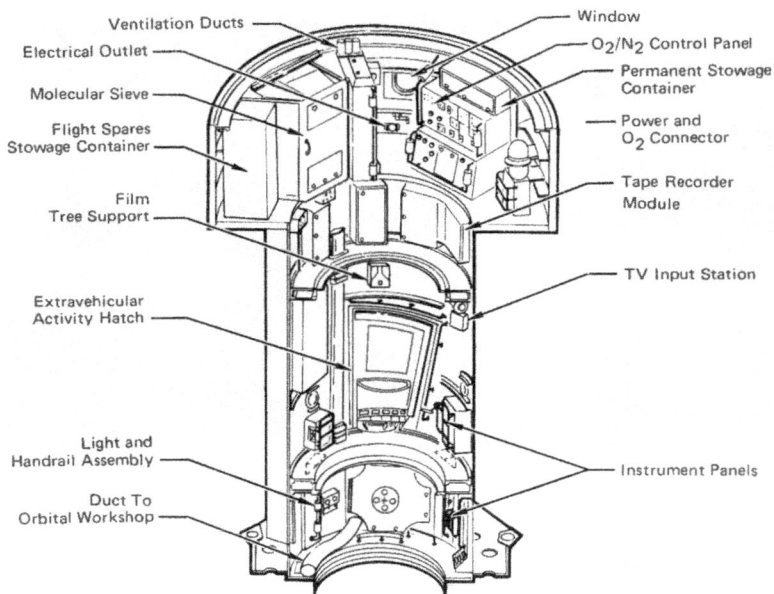

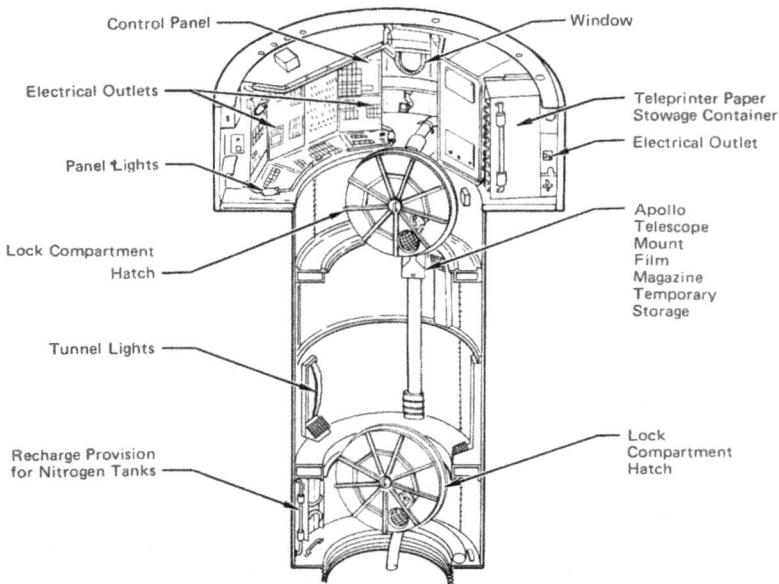

FIGURE 84.—Airlock Module with details of service and control panels.

74

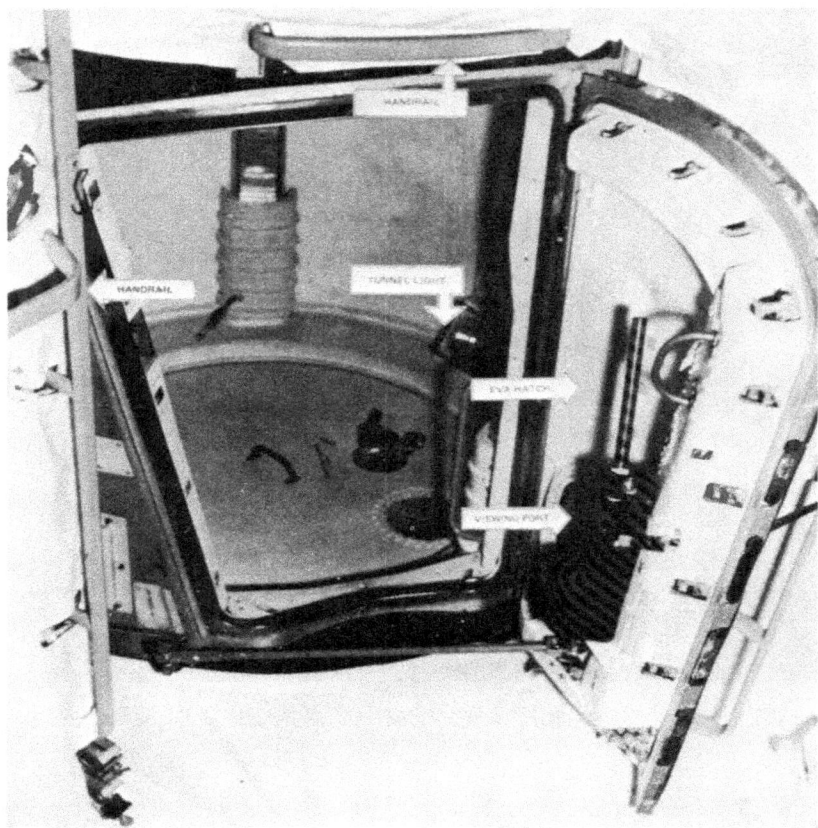

FIGURE 85.—Wall of Airlock Module, showing EVA hatch for exit and entrance during extra-vehicular activities.

will roll the canister by manually controlled switches until one of the cassette doors faces his EVA work station at the ATM rack (Fig. 94).

The telescope-bearing canister is not attached directly to the roll ring. Another concentric ring (gimbal ring), mounted between roll ring and canister and connected to the canister and to the roll ring in the fashion of a universal joint, will permit the canister axis to move relative to the rack axis around two perpendicular axes in order to achieve fine pointing and attitude stabilization of the telescopes. These angular motions need not be large because the Skylab cluster as a whole will always achieve coarse pointing. Maximum canister axis deflections relative to the rack axis will be $\pm 2°$. The bearings allowing these angular motions consist of flexure pivots (Fig. 95) which provide small angular deviations by flexing metal bands rather than by moving parts. Rotational motion of the canister around the two "universal joint" axes will be accomplished by electric torquers (Fig. 95) upon manual or automated signals.

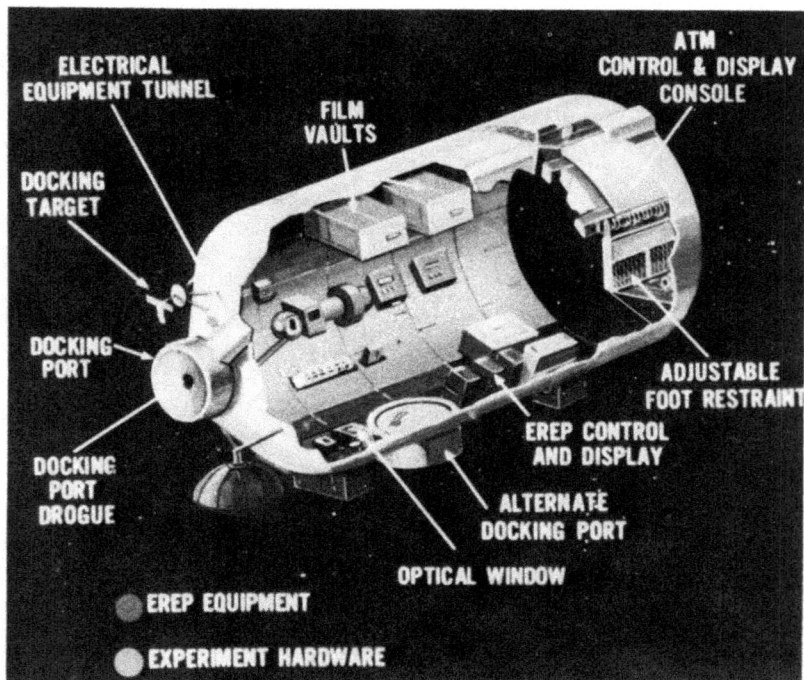

FIGURE 86.—Multiple Docking Adapter (MDA) with major components and installations.

A cylindrical shroud covers the experiment canister for protection against contamination and for thermal control purposes. The canister, 3 m (10 ft) long and 2.1 m (7 ft) in diameter, will maintain a steady temperature of 21° C (70° F) by means of an active cooling system which employs cooling pipes and a mixture of methanol and water to exchange heat between the canister wall and radiation cooling panels.

All eight solar telescopes, the fine Sun sensors, and some auxiliary systems are mounted on the "spar," a cruciform light-weight mounting panel which divides the canister lengthwise into four equal compartments (Fig. 96). The spar is designed for high rigidity; it serves as an optical bench for the solar telescopes. However, three of the experiments (S052, S055, and S082B) are built to apply additional individual fine corrections to their pointing directions. Each of the four compartments carries two of the solar telescopes (Fig. 97). The front end of the canister is covered by a sunshield with openings for telescopes and Sun sensors (Fig. 98). During ground storage and launch, these openings remain closed by individual doors for protection.

Operation of the solar telescopes and their auxiliary equipment will be monitored and controlled from the Control and Display Panel inside the Multiple Docking Adapter (Fig. 99). Experiment controls are located in the center of the panel, thermal control and lighting control readouts and switches are on the left, pointing control system displays and switches are

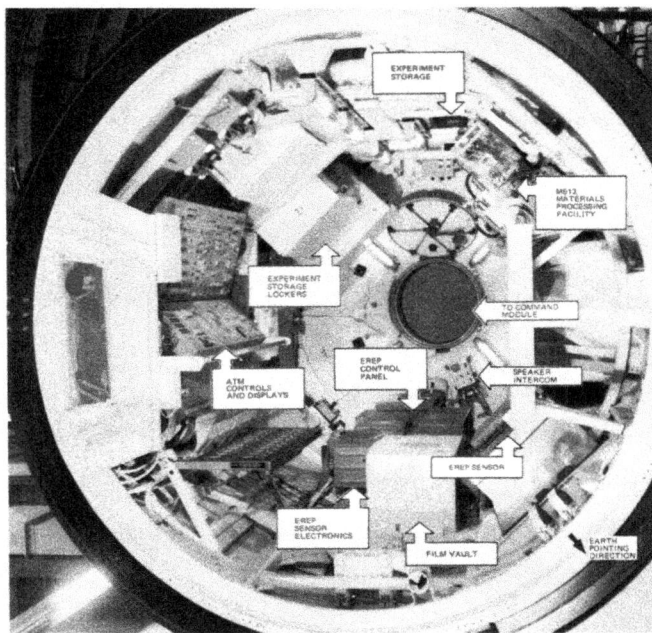

FIGURE 87.—Inside view of Multiple Docking Adapter with major components.

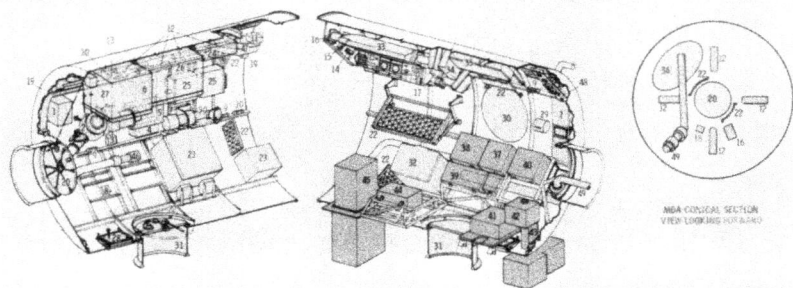

MDA CONICAL SECTION
VIEW LOOKING FORWARD

EXPERIMENTS
1. M479 SAMPLE CONTAINER
2. M518 DRYER
3. EXOTHERMIC SPECIMEN CONTAINER
4. COMPOSITE CASTING SAMPLE CONTAINER
5. ACCESSORIES CONTAINER
6. CRYSTAL SAMPLE CONTAINER
7. S009 EXPERIMENT

ELECTRONICS
8. WINDOW HEATER CONTROL
9. UTILITY OUTLET
10. HIGH POWER ACCESS OUTLET
11. RADIO NOISE BURST MONITOR
12. GMT
13. INTERCOM & UV-DETECTOR PANEL
14. UTILITY OUTLET & TV INPUT
15. VIDEO CONTROL
16. INTERCOM
17. DIGITAL ADDRESS SYSTEM
18. INTERIOR LIGHT SWITCH
19. UV DETECTOR

STRUCTURE
20. PRESSURE HATCH
21. FILM VAULT NO. 2
22. FOOT RESTRAINT/HANDRAIL
23. S79 MISC STORAGE CONTAINER
24. MISC STORAGE BOX
25. S082A AND S082B
26. FILM VAULT NO. 1
27. FILM VAULT NO. 4
28. MDA WINDOW
29. FIRE EXTINGUISHER
30. PROBE STOWAGE
31. DOCKING PORT
32. FILM VAULT NO. 3
33. ATM C&D CONSOLE
34. FLIGHT DATA FILE CONTAINER
35. C&D CONTAINER
36. DROGUE STOWAGE AREA

EREP
37. TAPE RECORDER
38. 10 BAND MULTISPECTRAL ELECTRONICS
39. S190 STORAGE CONTAINER
40. C&D PANEL
41. S190 EXPERIMENT
42. VIEWFINDER
43. VIEWFINDER ELECTRONICS
44. SPARE TAPE RECORDER
45. 10 BAND MULTISPECTRAL SCANNER

MECHANICAL
46. FAN NO. 1
47. ECS DUCT/FEDM HEAT EXCHANGER
48. VACUUM VENT PANEL
49. ECS FLEX DUCT
50. FAN NO. 2

FIGURE 88.—Multiple Docking Adapter, arrangement of experiments and structural elements.

FIGURE 89.—Skylab Shroud during separation in orbit.

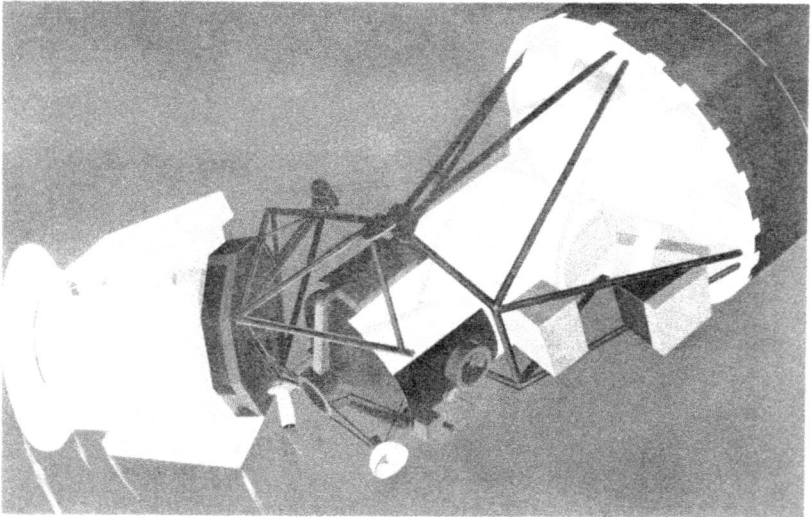

FIGURE 90.—Deployment of Apollo Telescope Mount (ATM) by rotation through 90 degrees.

on the upper right, power controls on the lower right and alert indicators above the center panels. Two TV screens on the panel will show selectively images from five different viewing instruments; they will enable the astronauts to select targets, to correct for pointing errors or drifts, to follow visually the development of active areas on the Sun, and to compare findings with observers on the ground by voice link.

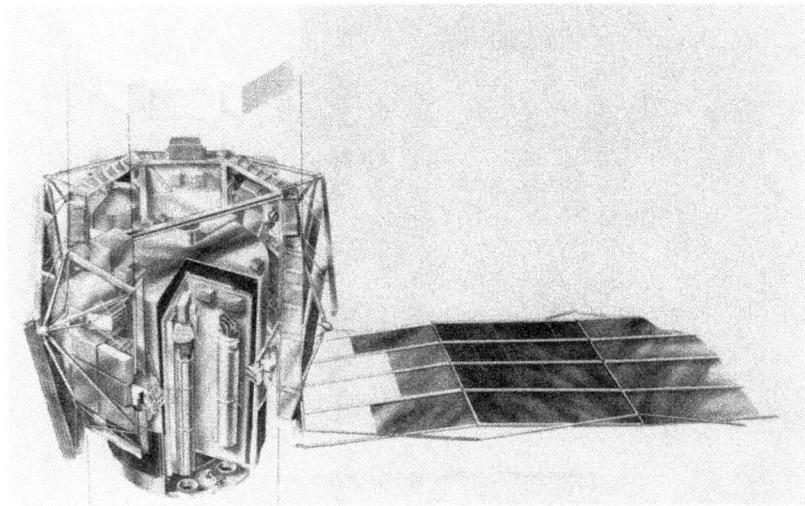

FIGURE 91.—Apollo Telescope Mount showing external rack with control moment gyroscopes and solar panels, and internal canister with telescopes.

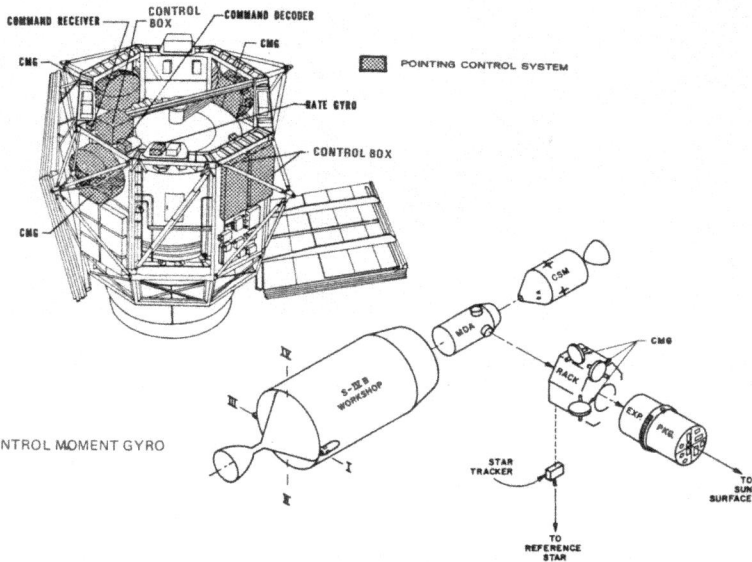

FIGURE 92.—Control moment gyroscopes, needed to provide fine attitude control for the Skylab cluster.

FIGURE 93.—Apollo Telescope Mount (ATM) during assembly, showing telescopes, roll ring, and inner gimbal system.

During launch and ascent, the Skylab cluster will be covered by the Payload Shroud (Fig. 100).

f. Command and Service Module (CSM)

The astronauts will be transported from the ground to Skylab, and back to the ground, by the Apollo Command and Service Module (Fig. 101). Basically, the CSMs for Skylab are identical with the CSMs used on the Apollo flights; the differences concern the capabilities of power and life support systems (Fig. 102). While the CSM on an Apollo mission had to be capable of sustaining its operation over a period of 14 days, the Skylab CSMs need to support their operations only for the short periods of ascent

FIGURE 94.—Apollo Telescope Mount (rack) with opening, showing canister inside the rack with a hatch to exchange film cassettes.

and descent. During the time of docking in orbit. CSM systems, which include data and communications systems, will be supported by the Skylab supply sources.

The Command Module, 4 m (13 ft) in diameter and 3.6 m (12 ft) high, contains a crew compartment for three astronauts, a docking tunnel to the top of the cone-shaped module, and a hatch that can be opened from the inside after docking with the Multiple Docking Adapter. Twelve latches at the outside of the tunnel end will attach the Command Module firmly to the port of the MDA before crew transfer can begin (Fig. 103).

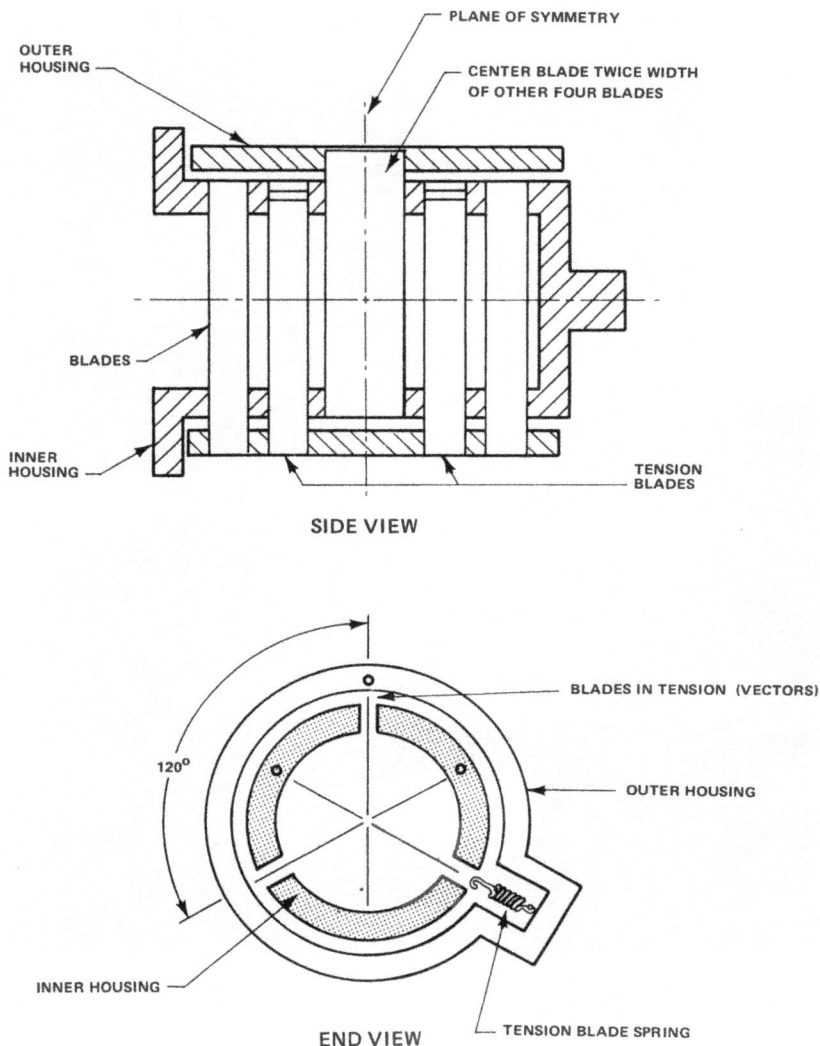

PLANE OF SYMMETRY

CENTER BLADE TWICE WIDTH
OF OTHER FOUR BLADES

OUTER
HOUSING

BLADES

INNER
HOUSING

TENSION
BLADES

SIDE VIEW

BLADES IN TENSION (VECTORS)

120°

OUTER HOUSING

INNER HOUSING

END VIEW

TENSION BLADE SPRING

FIGURE 95.—Flexure pivot bearings of the ATM canister, permitting fine attitude control.

Inside the Command Module are the guidance and control system, electric batteries, oxygen containers, control and display panels for the combined CSM, couches for the three crew members, stowage for the consumables needed during ascent and descent, and provisions for the stowage of equipment to be transported to or from Skylab. A heat shield, coated with ablative material, will protect the Command Module against the heat produced during reentry (Fig. 104). Except for the last half hour during reentry, the Service Module will remain firmly attached to the Command Module. It contains

FIGURE 96.—ATM canister, showing gimbal rings and the internal spar which carries the telescopes.

service systems and supplies that do not require access by crew members during flight, such as the main propulsion system, a maneuvering system, fuel cells for electric power, part of the oxygen for breathing, and radiators for cooling. After separation from the Command Module shortly before reentry, the Service Module will heat up in the atmosphere and burn completely; only ashes will reach the ground.

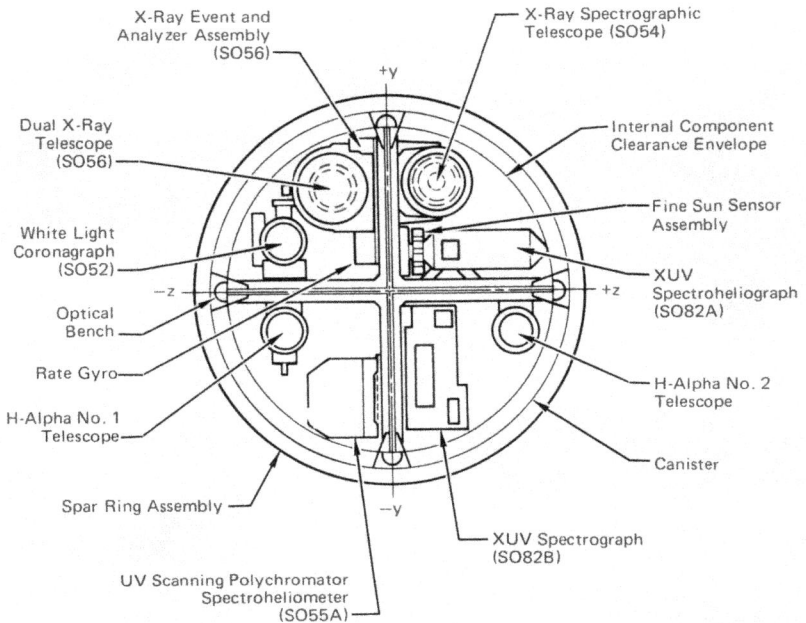

X-Ray Event and Analyzer Assembly (SO56)

X-Ray Spectrographic Telescope (SO54)

+y

Dual X-Ray Telescope (SO56)

Internal Component Clearance Envelope

Fine Sun Sensor Assembly

White Light Coronagraph (SO52)

−z

+z

XUV Spectroheliograph (SO82A)

Optical Bench

Rate Gyro

H-Alpha No. 2 Telescope

H-Alpha No. 1 Telescope

Canister

Spar Ring Assembly

−y

XUV Spectrograph (SO82B)

UV Scanning Polychromator Spectroheliometer (SO55A)

Telescope Installations

FIGURE 97.—Cross section through ATM canister, showing the cruciform spar and the telescopes.

g. Integration, Testing, and Quality Assurance

All the elements of Skylab components produced at many different plants and places were subjected to elaborate procedures of component testing and integration. Tests had to be performed on individual elements, on subassemblies of elements, and finally on the completed assemblies. Some tests were performed at the manufacturing sites; others were performed at assembly sites. Fig. 105 illustrates how the major elements of Skylab flight components flowed together in the testing and integration operations.

The Orbital Workshop was assembled with experiments, thruster attitude control system, and habitability support systems at Huntington Beach, California. It was then shipped to Kennedy Space Center (KSC) for test, checkout, and assembly together with other Skylab elements in the Vertical Assembly Building (VAB). Before being transported to the launch pad, it underwent an integrated systems test with all of the other modules except the CSM.

The Apollo Telescope Mount experiments were assembled in the ATM canister at MSFC. After vibration testing, the assembly was first shipped to JSC for a thermal vacuum test, and from there to KSC for systems test and

FIGURE 98.—Apollo Telescope Mount front end at the moment of film retrieval by astronaut. Telescope orifices are closed.

checkout. It was then assembled with other Skylab elements, tested as a major component of the total system, and transported to the launch pad.

The Payload Shroud was manufactured on the west coast, and then shipped to KSC for assembly, integrated systems test, and launch.

The Airlock Module was built and fitted with experiments in St. Louis, Missouri.

The Multiple-Docking Adapter shell was built at MSFC, shipped to Denver, Colorado, where it was outfitted and tested, and then sent to St. Louis for assembly with the Airlock Module. The AM–MDA assembly then underwent systems tests, altitude chamber tests, and checkout before it was shipped to KSC for integration with the CSM and testing.

The AM–MDA was then moved to the VAB, assembled with the other payload modules, tested as part of an integrated systems test on top of the launch vehicle, and transported to the launch pad for final checkout and launch.

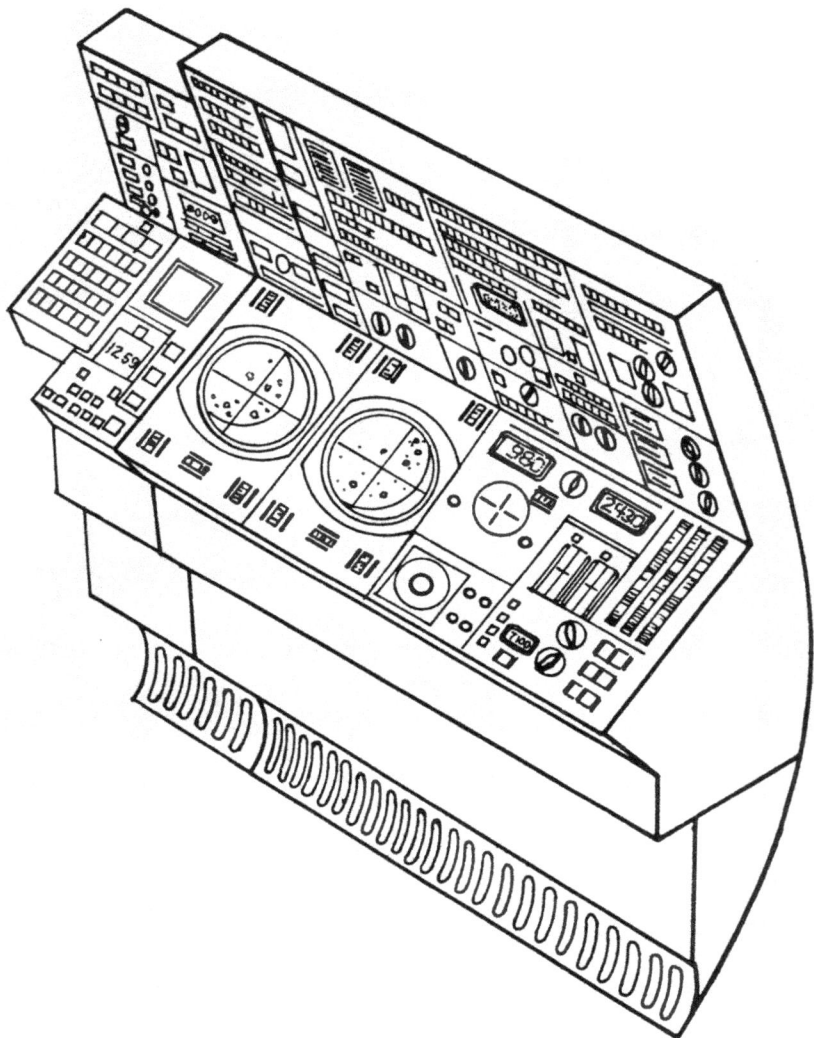

FIGURE 99.—Console for monitoring and controlling the telescopes in ATM.

From the beginning of Skylab, the desire for a high probability of success has made it necessary that careful attention be given to all aspects of reliability and quality assurance. Experience gained in many previous space projects formed the basis for a comprehensive reliability program established specifically for Skylab. Quality and reliability requirements were carefully documented in numerous plans which controlled all phases of the Skylab from design through production, test, launching, and operation in space. These reliability requirements were imposed uniformly upon all contractors and suppliers who participated in Project Skylab.

FIGURE 100.—Payload shroud, needed to protect the front part of Skylab during ascent.

The specific quality and reliability program requirements generated for the Skylab project required that only such parts could be used for Skylab which either had been flight proven or had undergone elaborate and rigorous testing. Likewise, only such vendors were accepted who either had records of successful performance on previous flights or had been found acceptable after very careful screening. An attempt was made to use for Skylab as many components of previous successful flight projects as possible; if new components had to be developed, they were subjected to extensive testing. The principle of high reliability was made a basic feature of design from the very beginning of Skylab program work. A thorough failure-mode analysis had

87

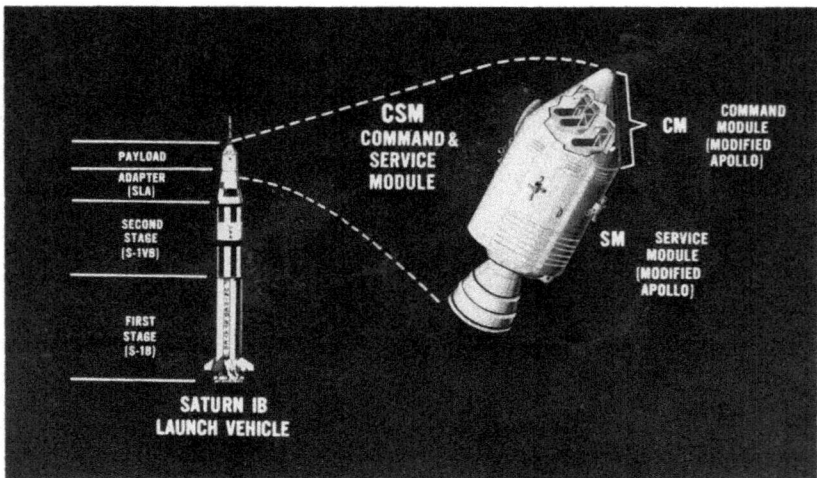

FIGURE 101.—Command and Service Module (CSM) launch configuration.

to be made of every functional part of the complex Skylab system. If it was found that failure of a single component or system could jeopardize the mission, the design had to be changed, or redundancy [1] had to be incorporated; if neither was feasible, a specific effort was made in the manufacturing and testing of that component or system to assure the highest reasonable degree of reliability. As part of the test program, an elaborate failure reporting system was established. If a failure occurred, its cause was traced back to its origin, and the likelihood of a recurrence of a similar failure was eliminated by proper actions.

Experience has shown that a major step toward high quality can be accomplished by proper "procurement control." Drawings and specification lists must reflect the quality requirements to the extent that the vendor knows exactly what he is expected to deliver, and acceptance testing must be sufficiently rigorous to eliminate all weak parts and components.

2. SKYLAB OPERATIONAL SYSTEMS

Astronauts and experiments on Skylab are supported by a number of auxiliary or "housekeeping" systems for all those functions which are necessary to make Skylab a self-contained, well-functioning, and efficient laboratory. These operational systems include attitude control, environmental control, communications, data management, crew accommodations, and electric power generation.

Details of these systems are described in the following sections.

---

[1] Redundancy: See Glossary for explanation.

88

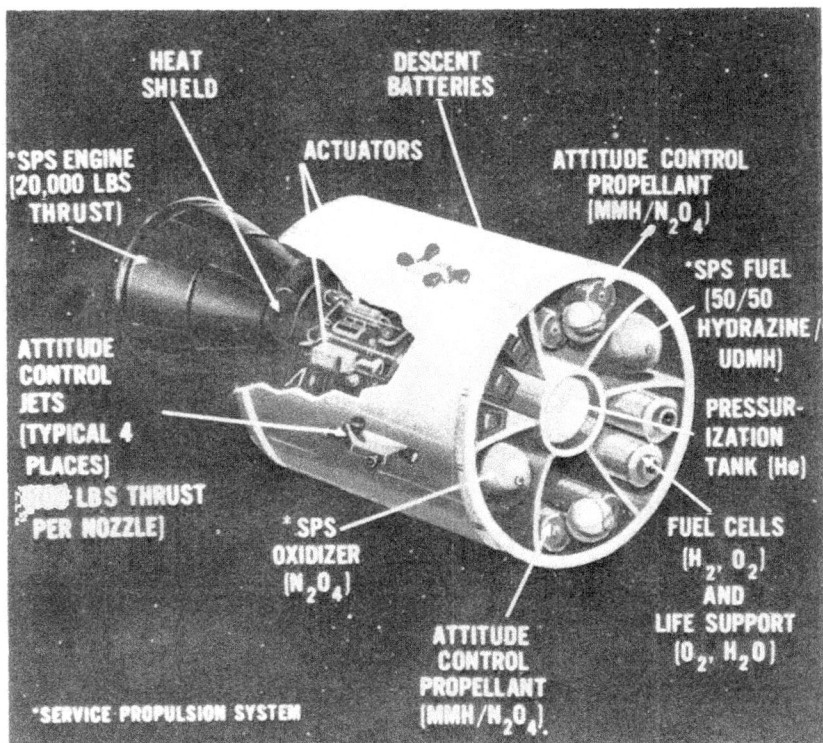

FIGURE 102.—Service Module showing major components and equipment. (MMH is monomethyl hydrazine, N_2O_4 is nitrogen tetroxide, UDMH is unsymmetrical dimethylhydrazine, He is helium).

a. Attitude and Pointing Control System

Skylab experiments will view three basic targets; the Sun, the Earth, and celestial space. Instruments for these experiments are located at places which will provide proper viewing directions with a minimum of maneuvering (Fig. 106). Active and continuous control of Skylab attitude will assure that instruments are pointed in their desired directions during the periods of their operation.

The solar power arrays on the Workshop and the Apollo Telescope Mount will require orientation toward, or at least nearly toward, the Sun for as much time as possible.

Pointing accuracy requirements of the ATM solar telescopes around the three axes of Skylab, illustrated in Fig. 107, are listed below.

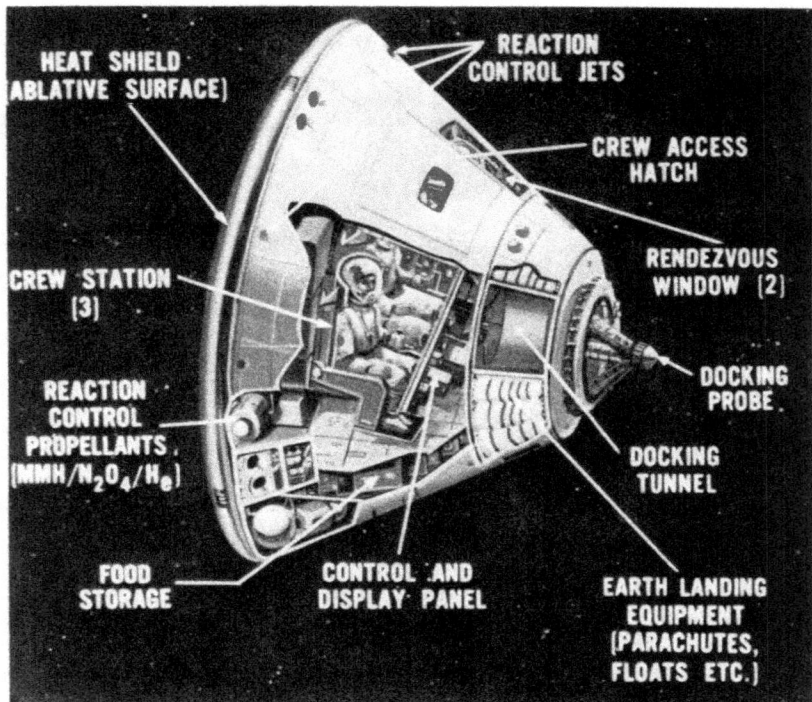

FIGURE 103.—Command Module with major components and installations.

FIGURE 104.—Command Module with parachutes shortly before splashdown.

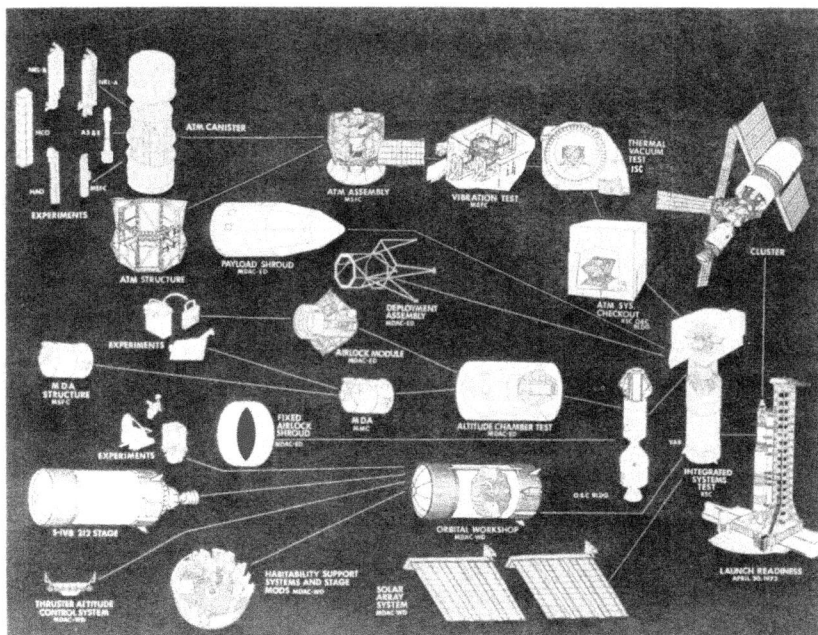

FIGURE 105.—Flow of Skylab components from manufacturing to testing, integration, and launch.

TABLE 5.—*Pointing Accuracy of ATM Solar Telescope Canister*

System axis	Pointing accuracy	Stability
X	±2.5 arc sec	±2.5 arc sec/15 min.
Y	±2.5 arc sec	±2.5 arc sec/15 min.
Z	±10 arc min	±7.5 arc min/15 min.

These pointing accuracies for the telescopes can be accomplished because of the gimbal mounting of the ATM experiment canister within the ATM rack. Pointing accuracies to which the entire Skylab cluster can be held in various observational modes are listed below.

TABLE 6.—*Pointing Accuracy of Entire Skylab Cluster*

System axis	OBSERVATION MODE		
	Solar	Earth or sky	During docking maneuvers
X	±6 arc min	±2 degree	±6 degree.
Y	±6 arc min	±2 degree	±12 degree.
Z	±10 arc min	±2 degree	±6 degree.

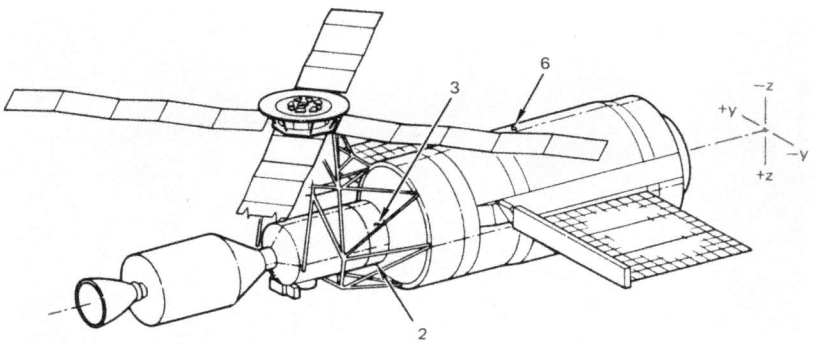

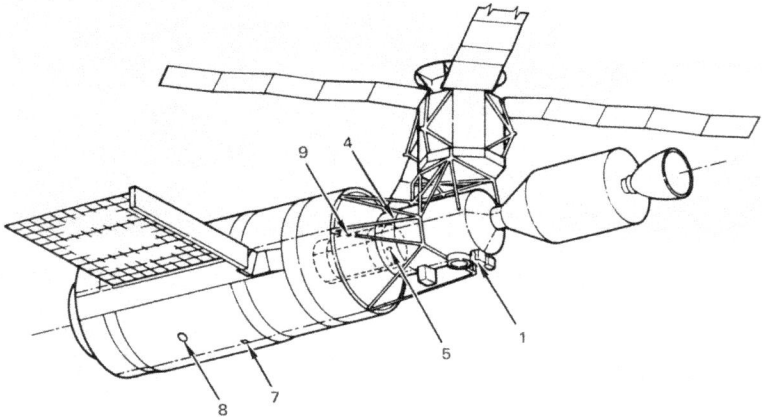

Port Angular Locations

Port No.	Window Name	Location
1.	S190 Experiment Window	On +z Axis
2.	Airlock Module Window	37.5 deg Off +z Toward −y
3.	Airlock Module Window	37.5 deg Off −y Toward −z
4.	Airlock Module Window	37.5 deg Off −z Toward +y
5.	Airlock Module Window	37.5 deg Off +y Toward +z
6.	Solar Scientific Airlock	0.14 deg Off −z Toward +y
7.	Antisolar Scientific Airlock	4.3 deg Off +z Toward −y
8.	Wardroom Window	25.9 deg Off +z Toward +y
9.	EVA Hatch Window	45 deg Off −z Toward +y

FIGURE 106.—Viewing ports on Skylab.

Controlling the attitude of Skylab will be the task of the Attitude and Pointing Control System. This function will include rotating the Skylab cluster to the desired orientation, holding this orientation as long as necessary, and providing the high precision pointing control for the Apollo Telescope Mount. In order to execute these actions, the Attitude and Pointing

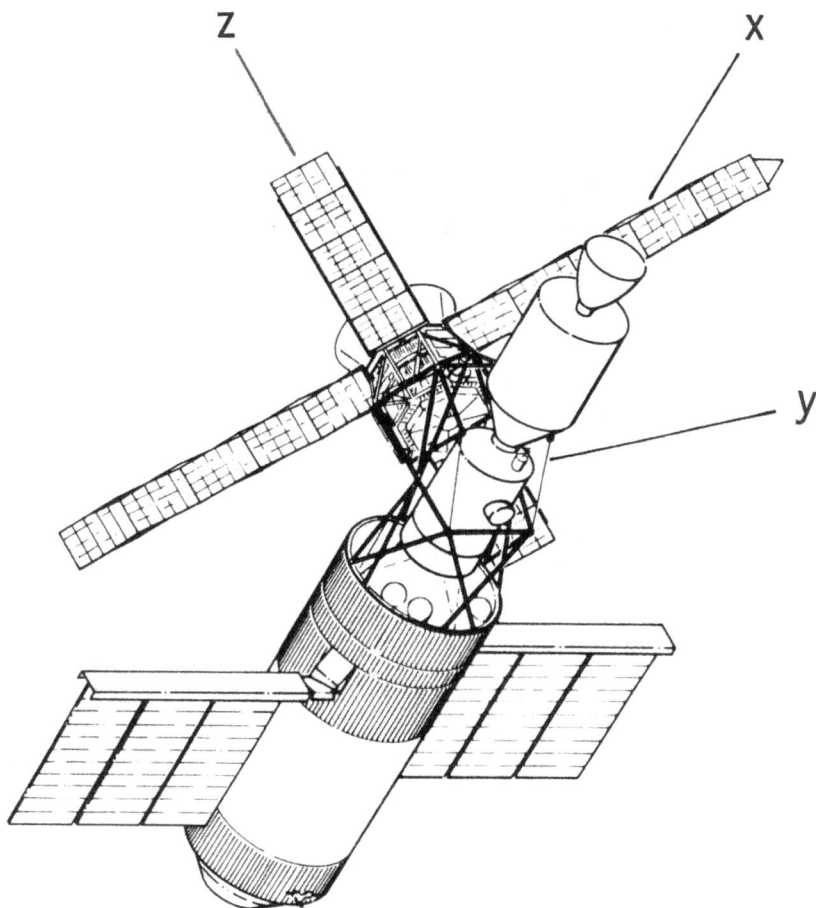

FIGURE 107.—Principal axes of Skylab cluster.

Control System uses sensors to read out the existing attitude with respect to reference directions and a mechanism to change the attitude in a controlled fashion.

Skylab will use rate gyroscopes as basic sensing elements for its attitude control system. Rate gyroscopes measure the rate of angular rotation of Skylab around each of the three principal axes. By intergrating these angular rates over a given time, the angular changes during this period will be obtained. Reference directions from which angular changes can be counted will be provided by Sun and star seekers (Figs. 110, 111). The Sun seeker which monitors the solar reference direction will aim at the center of the solar disc. The star seeker will aim at one of three stars, preferably at Canopus in the southern constellation Argus.

FIGURE 110.—Sun seeker (acquisition sensor) for angular guidance.

Changes in Skylab attitude can be accomplished by either one of two operational control systems, the Control Moment Gyro System (CMG) or the Nitrogen Thruster Attitude Control System (TACS). The CMG control system, consisting of three large gyroscopes with mutually perpendicular axes represents the prime method of Skylab control (Fig. 112). The TACS system can also control Skylab in a manner characteristic of conventional cold gas thruster control systems (Fig. 113). In these systems, jets of gas are released through rocket-type nozzles, thus producing small amounts of thrust.

Each of the three CMGs in the Skylab control system weighs 181 kg (400 lbs.); it consists of a rotor of 0.55 m (22 in.) diameter, spinning at 9,000 RPM, and an inner and an outer gimbal ring. The outer gimbal ring permits Skylab to rotate around the gimbal axis of each CMG. However, an electric torque motor attached to Skylab and acting upon the outer gimbal ring can produce a torque between Skylab and that CMG, resulting in a tilting motion (precession) of the CMG rotor axis around the inner gimbal ring axis by virtue of the characteristic property of a spinning gyroscope to respond to torques around one axis with a tilting motion (precession) around the other axis. The reactive force of the torque motor then causes Skylab to change its angular position while the rotor axis moves (precesses). Operation of the torque motor is controlled by commands received from the Skylab digital control computer. During an attitude control procedure, the torque motors of all three CMGs will normally receive control commands,

94

FIGURE 111.—Star seeker for angular guidance.

and each CMG rotor axis will slowly tilt. Should the control torque persist long enough, the tilting motions would continue until the rotor axis of each gyro eventually would be parallel with the axis of the control torque. From that moment on, none of the CMGs could continue to react to the control torque of the torque motor by further precession. If all three CMG rotor axes should be parallel with their torque axes, the CMG system would be "saturated" and would no longer be capable of controlling the attitude of Skylab. In order to avoid this saturation, a desaturation procedure is provided which utilizes the gravity field of the Earth as generator of a counter-acting torque.

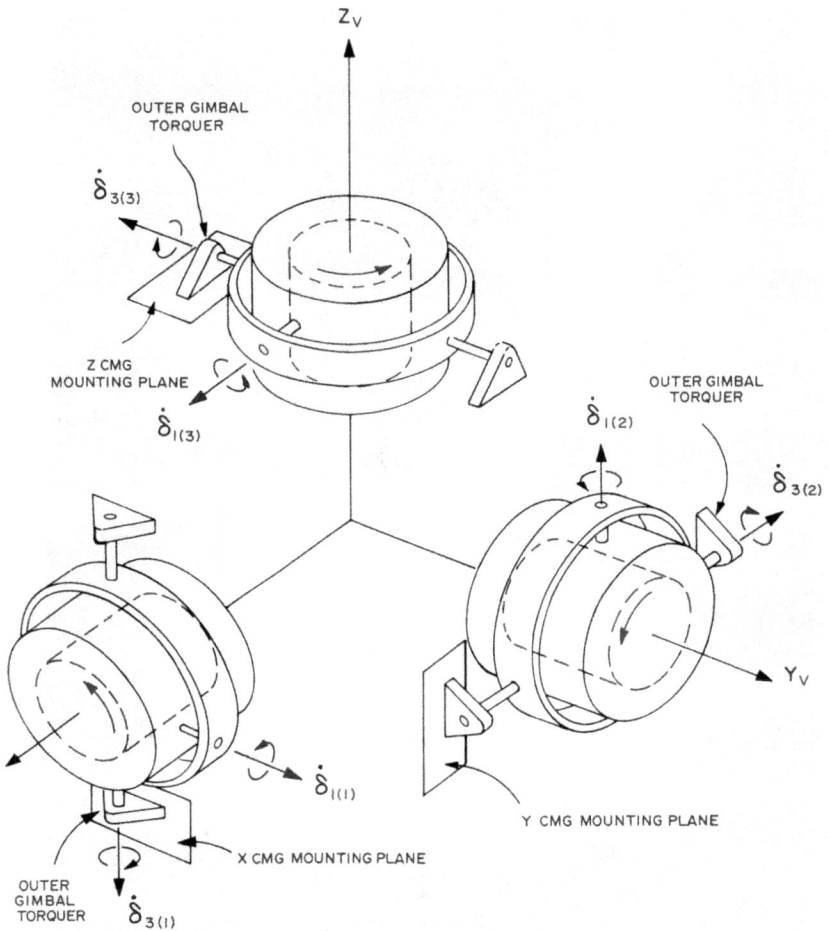

FIGURE 112.—Control scheme of three control moment gyroscopes.

On an orbiting spacecraft, gravitational forces are exactly balanced by centrifugal forces at every point on a line representing the orbital trajectory of the center of gravity of the spacecraft. Points below this line experience an excess of gravitational forces; points above this line, an excess of centrifugal forces. The amount of each of these forces at a given point on the spacecraft is a function of the distance between the point and the center-of-gravity line. In general, the integrated effect of these two forces represents a force couple, or a torque, called gravity gradient torque. Its magnitude depends, among other factors, on the shape and the attitude of the spacecraft. On Skylab, this gravity gradient torque is used as desaturating torque for the CMGs. For this purpose, a computer routine is incorporated in the Attitude and Pointing Control System which continuously calculates the amount of desaturation required. On the night side of each orbit when the

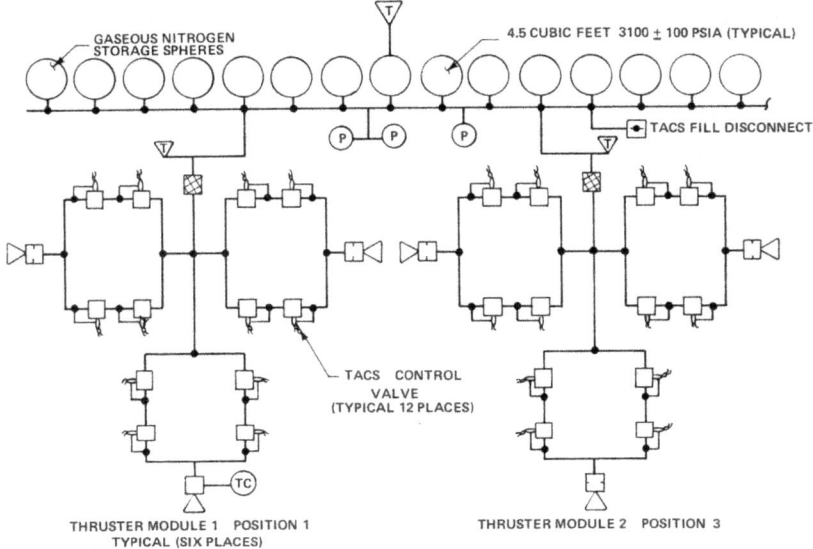

FIGURE 113.—Reaction nozzles for angular control, using compressed nitrogen. Parallel and series valves for each nozzle increase reliability through redundancy.

Sun is not visible, proper signals are generated for the CMGs which turn the Skylab cluster into an attitude that produces an appropriate gravity gradient torque. The Attitude and Pointing Control System, in an effort to maintain this attitude, generates control torques against the gravity gradient torque in such a manner that the precession axes of the CMGs tilt until their previously accumulated precession angles are completely used up. This desaturation maneuver normally will suffice to restore the capability of the CMGs to fully control Skylab. Should the amounts of accumulated precession angles exceed the amounts that can be neutralized by the gravity gradient torque, the TACS will be energized to neutralize the difference.

It is anticipated, though, that desaturation of the CMGs by TACS operation will occur very rarely. This will help keep the external environment of Skylab reasonably free from gases and other contaminants.

During the first 7.5 hours after orbit insertion, Skylab will carry out several directional maneuvers (Fig. 20). Gyro platform and digital computers in the Instrument Unit (see Chapter IV–1–b) will generate the necessary control signals, and the thruster attitude control system will execute the maneuvers. After this initial period, the Instrument Unit will transfer the control authority to the ATM digital computer which will use the Control Moment Gyros as the prime system for attitude control and the thruster system only when needed for CMG desaturation. This combined system will suffice for the accuracy requirements of Skylab as listed in Table 6. Higher accuracies as required by the solar telescope canister in ATM (Table 5), will be provided by the Experiment Pointing Control Loop which consists of the canister flexure pivots and associated torque motors. Signals for the attitude control of the ATM canister are derived from fine Sun sensors and rate gyros on the canister and processed in the Experiment Pointing Electronic Assembly. By overriding these signals, the astronauts can move the ATM canister to a specific target on the Sun with a hand controller from the ATM Control Console in the MDA (Fig. 99). At any selected position, the Experiment Pointing Control Loop system will keep the pointing accuracy and stability within the necessary limits.

b. Environmental Control System

Environmental control within Skylab will be achieved by an open-cycle life support system in which the consumables are not reclaimed for reuse. Before each crew arrives, Skylab will be pressurized to 34,000 Nm^{-2} (one-third of an atmosphere or 5 psi) of a mixture of oxygen and nitrogen (approximately 74 percent oxygen and 26 percent nitrogen). This mixture will assure that the astronauts will breathe the same amount of oxygen as they would on Earth. The Airlock Module, serving as a nerve center for the total Skylab cluster, among other functions will also control the internal atmosphere and temperature. The atmospheric gases will be stored in gaseous form in bottles mounted on the Airlock Module (Fig. 8). Flow regulator valves will ensure that correct pressure and gas mixture are maintained.

Relative humidity will be controlled to about 26 percent at 30° C (86° F). The carbon dioxide concentration will be held below a maximum pressure level of 700 Nm^{-2} (7 millibars). Temperature in the habitation areas will be controlled between 13° and 32° C (55° and 90° F).

Life support and atmospheric consumables will include 2,700 kg (6,000 lbs) of water, 670 kg (1,470 lbs) of food, 2,240 kg (4,930 lbs) of oxygen, and 600 kg (1,320 lbs) of nitrogen. Environmental operating limits are shown in Table 7.

Skylab systems have been designed to ensure that at any given time the overall sound pressure level will not exceed 72.5 db above the normal human hearing threshold [1] when the individual sound pressure levels from all sources are added together.

Atmospheric purification and humidity control will be achieved by passing the cabin gases through carbon dioxide removal equipment and through water removal condensers. Odors will be removed by passing the gases through charcoal filters (activated coconut shell charcoal).

The Skylab CO_2 removal equipment consists of two units, each containing two beds of zeolite for absorption of carbon dioxide (Fig. 115). The beds will operate on a reversible cycle. After absorbing CO_2 for 15 minutes, a bed will undergo a process that extracts most of the CO_2 and expels it overboard; during this purging time, the other bed will be switched to the absorbing mode.

The thermal effects of solar irradiation on the daylight side and lack of any heat source on the dark side have been almost eliminated as influences on the temperature inside Skylab through insulation and thermal coatings around the Workshop, the Airlock Module, and the Multiple Docking Adapter.

---

[1] A sound level of 72.5 db is approximately equal to that of a "noisy office."

TABLE 7.—*Environmental Conditions in Crew Quarters*

PERIOD	Temperature		Relative humidity %	Pressure [1]			Gas
	°C	°F		Nm⁻²	psi	Atm.	
Prelaunch non-operational.	−18 to 71	0 to 160	30 to 45..	1.2 x 10⁵ to 1.8 x 10⁵	16.8 to 26.5	1.2 to 1.8	Air.
Prelaunch operational.	5 to 27	40 to 80	0 to 40...	10⁵ to 1.8 x 10⁵	14.7 to 26.5	1 to 1.8	20 to 0% O₂. 80 to 100% N₂.
Launch and ascent.	5 to 43	40 to 110	...do.....	1.6 x 10⁵ to 1.8 x 10⁵	23.5 to 26.5	1.6 to 1.8	100% N₂.
Orbit operational..	13 to 32	55 to 90	25 to 85..	0.33 x 10⁵ to 0.35 x 10⁵	4.8 to 5.2	0.33 to 0.35	74% O₂. 26% N₂.
Orbit storage......	5 to 30	40 to 85	25 to 100.	0.03 x 10⁵ to 0.4 x 10⁵	0.45 to 6	0.03 to 0.4	74% O₂. 26% N₂.

[1] Nm^{-2} = Newtons per square meter.
psi = pounds per square inch.
Atm = atmosphere.

To control temperatures and humidity during manned and unmanned periods, an active thermal control system for OWS, AM, and MDA has been provided. Located in the Airlock Module, this system will cool and purify the atmosphere. A combination of air duct heaters and wall heaters, located in other areas of the Skylab, will provide heating as required. The heaters will prevent condensation from forming and damaging instruments and equipment, and they will maintain a comfortable environment for the crews. Some heat-producing components that require cooling are mounted on cold plates which will be temperature-controlled by a liquid coolant system in the AM. Excess heat from the AM cooling system will be radiated to space through radiators on the MDA and on the forward AM.

Thermal control of the Apollo Telescope Mount will be provided by a system of passive control measures, radiant heaters, and a liquid coolant system with cold plates and space radiators. A Sun shield will protect much of the ATM equipment from direct sunlight (Fig. 98).

c. Data and Communications Systems

A space system as active and complex as Skylab produces a huge amount of data which must be sent to Earth for evaluation and use. Basically, there are two types of data; first, physical data including film, tapes, emulsion plates, surface samples, biomedical specimens, log books, and notes; and

FIGURE 115.—Carbon dioxide removal system, using reusable zeolite beds.

second, data in the form of signals such as audio, telemetry, and video signals. Physical data will be brought back from Skylab on the Command Module. Transmittable data will be sent from Skylab to Earth, and also from the Earth to Skylab, through a system of transmitters and receivers on Skylab and in the Spacecraft Tracking and Data Network (STDN).

Fig. 116 shows the locations and the approximate ranges of 13 STDN stations, 11 of them fixed, one ship-borne, and one air-borne. The Skylab communications system, whose characteristics are listed in Table 8, provides the links between Skylab and the STDN. In addition to real-time telemetry, which will be available during about one-fourth of the time with an average contact time of 6.5 minutes per station, delayed time data and voice will be recorded on board for playback while Skylab is over a ground station. The playback system has the capability to dump two hours of stored data in 5.45 minutes. Periodic television transmission for the five ATM cameras and for the portable TV cameras will be achieved through the frequency-modulated S-band link on the Command and Service Module. A video tape recorder system will be available in the Multiple Docking Adapter. The portable color cameras, either hand-held or bracket-mounted (Fig. 118), have 525 scan lines at 30 frames per second. They can operate within a wide range of illumination.

Some of the scientific data, such as solar ultraviolet spectral measurements of Experiment S055, will be recorded on tape and transmitted to ground when Skylab is in line of sight for one of the receiving stations. Measurements taken by the numerous control instruments on Skylab, including housekeeping information on temperature, pressure, and humidity, and biomedical data from sensors worn by the astronauts, will be processed by

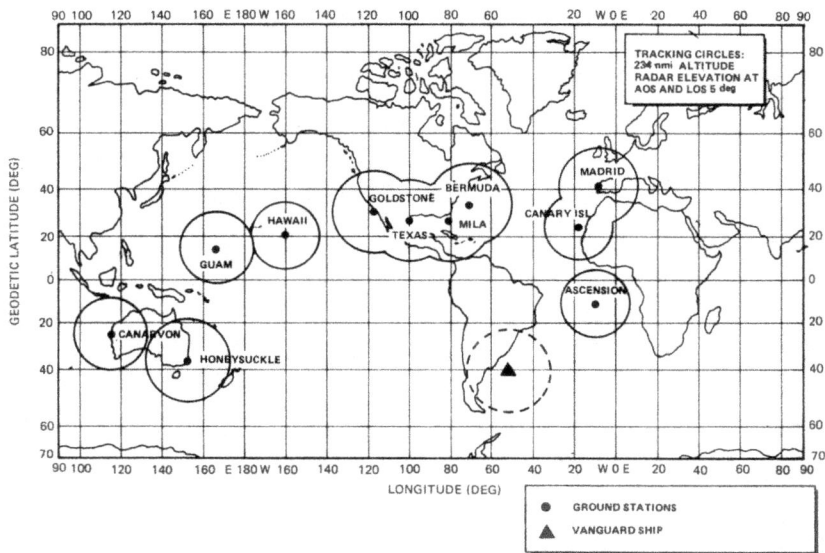

FIGURE 116.—Network of ground stations around the Earth for communications to and from Skylab.

onboard instrumentation systems and either transmitted directly to Earth or recorded on tape for transmission when ground contact exists.

Commands can be sent to Skylab from Mission Control Center in Houston to perform onboard functions, and to supply data to the onboard computers and the crew. An onboard teleprinter is available to produce paper copies of information for the crew, along with daily schedules of Skylab activities. Voice communication between Skylab and Mission Control at the Lyndon B. Johnson Space Center in Houston will be handled through the Command and Service Module docked at Skylab while astronauts are in orbit. Internal communications between astronauts anywhere in the Skylab cluster or on extravehicular activities will be possible through a number of communications panels and loudspeakers.

Principal Investigators (PIs) will have the opportunity of communicating with crew members through the Mission Control Flight Director.

Flow and distribution of Skylab data are illustrated in Fig. 119. All the signals received at STDN stations will first be passed on to the Goddard Space Flight Center, and from there to the Mission Control Center in Houston. The Principal Investigators will receive their data through Mission Control either in the form of tapes of processed and printed data, and of photographs as obtained from transmitted signals, or as films, samples, specimens, notes, and log books brought back in the Command Modules. After the Principal Investigators have analyzed the data and had an opportunity to publish the results, NASA will make the experiment data available to other qualified investigators upon request.

TABLE 8.—*Skylab External Communications*

Frequency MHz	Mode [1]	Modulation	Use
230.4	T	FM/PCM	Launch telemetry.
230.4	T	FM/PCM	Telemetry, voice, data.
235.0	T	FM/PCM	Telemetry, voice, data.
246.3	T	FM/PCM	Telemetry, voice, data.
450.0	R	FM	Ground Command teleprinter.
296.8	T	AM	CSM ranging.
259.7	R	Tones	CSM ranging.
231.9	T	FM/PCM	ATM telemetry.
237.0	T	FM/PCM	ATM telemetry.
450.0	R	FM	ATM ground command.
243.0	T	ICW	Recovery beacon.
259.7	T	AM	Ranging to Skylab.
259.7	R	AM	Ranging to Skylab.
296.8	T	AM	Voice, range, data.
296.8	R	AM	Voice, range, data.
2106.4	T	PM/PCM	Telemetry.
2287.5	T	PM/PCM	Telemetry.
2272.5	T	FM	TV, telemetry.

[1] T=Transmitter, R=Receiver.

FIGURE 118.—Handheld TV camera for use on Skylab.

The supply of photographic films for the complete Skylab mission includes about 280 cassettes, most of them with 400 ft-reels of 16 mm film; 64 magazines with 16 mm, 35 mm, or 70 mm film; and a number of packs and rolls with films of special sizes and varying lengths.

Weights of films, tapes, specimens, and other physical data expected to be brought back from Skylab in the three Command Modules are listed in Table 9.

d. Crew Accommodations

As a manned space vehicle, Skylab provides a habitable environment with living quarters, crew provisions, and facilities for food preparation and waste disposal to support a three-man crew for three missions (one for 28 days, and two for 56 days each). The center of activity is the Workshop shown in Fig 72. The main "floor" of the Workshop contains the wardroom or food management subsystem, individual sleeping compartments, an experiment work area, and the waste management subsystem (toilet) with hygiene facilities (Fig. 121).

The Food Management Subsystem consists of equipment and supplies required for the storage, preparation, and consumption of the daily meals by the astronaut crew. The astronauts, provided with a 140-day supply of food and beverages, will use the wardroom as kitchen and dining room. Food is stored in food boxes, galley trays, food freezers, and a food chiller. A galley, the food table, food trays, and eating utensils are provided for the preparation and consumption of the meals.

For the first manned space flights during the early 1960's, meals were prepared and packaged specifically for the zero-gravity environment with the result that they barely resembled normal Earth food. Real progress in astronaut eating was achieved on Christmas Day, 1968, when Frank Borman, James Lovell, and William Anders, circling the Moon on board Apollo 8, opened a surprise food package. It contained natural pieces of turkey with brown gravy, bright red cranberry applesauce, and a normal spoon. That meal proved that it was possible to serve tasty, familiar food, and to eat it with normal utensils, under zero gravity. Loose food and liquids could be controlled easily; even the gravy stayed where it belonged. A further step toward more normal food has been taken for Skylab. The Skylab crew will have a wide variety of frozen and dehydrated food to prepare menus ranging from cold cereals to potato salad, shrimp cocktail, and filet mignon. Foods that will adhere to a fork or knife, such as steak, mashed potatoes, or pie, will be eaten with normal utensils; liquids, including coffee, tea, instant breakfast, grape and orange drink, cocoa, and lemonade, will be served in squeezable plastic containers and sipped through tubes.

The galley in the wardroom will provide the daily supply of food; galley-located equipment will be used for preparation and disposal of food. Meal preparation and consumption equipment is shown in Fig. 122; trash disposal in Fig. 123.

The food table will permit three crewmen to simultaneously heat their food and to eat their meals in an efficient and comfortable manner, using

103

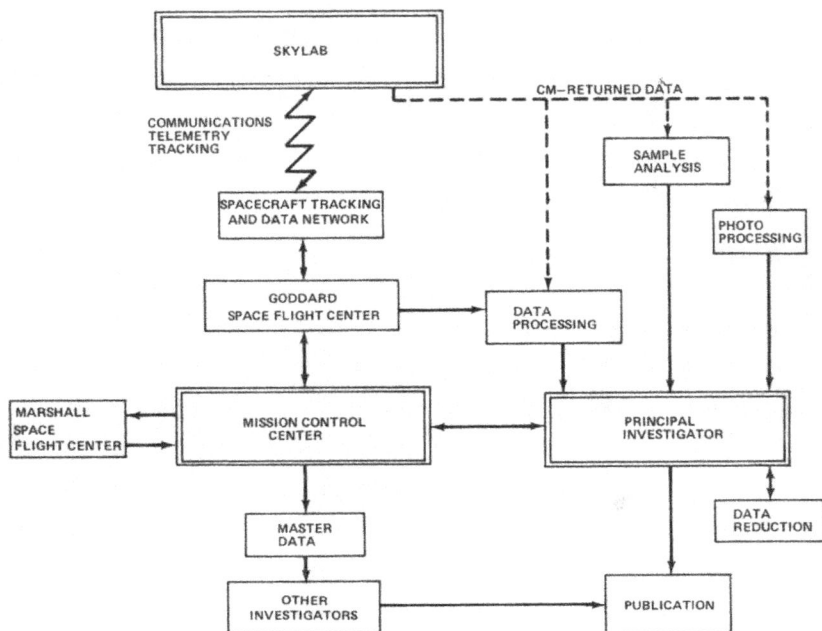

FIGURE 119.—Flow and distribution of data returned from Skylab.

TABLE 9.—*Weights of Data and Material to Be Returned On Board Command Modules*

Type of material	Mission					
	I		II		III	
	kg	lbs	kg	lbs	kg	lbs
Biomedical Specimens.....................	36	79	48	106	46	101
Solar Astronomy Films...................	84	185	168	370	84	185
Science Films and Samples...............	34	75	16	35	5. 5	12
Earth Resources Films and Tapes..........	35	77	35	77	40	88
Technology Films and Samples............	30	66	2	4	....	....
Operations Films........................	24	53	28	62	13	27

normal utensils and food trays. The table also supports components of the water system, including the water chiller and the wardroom water heater. The water chiller provides cold water for cold reconstitution of dehydrated foods and beverages and also for drinking purposes. The wardroom water heater provides hot water for hot reconstitution of dehydrated

SKYLAB WORKSHOP
CREW QUARTERS

FIGURE 121.—Main floor of Skylab, showing hatch for waste disposal (formerly oxygen tank) in foreground, and waste management subsystem (toilet) in rear center.

FIGURE. 122.—Galley equipment, including food table and food storage lockers.

foods and beverages. Each eating station has a foot and thigh restraint to hold the crewman in a comfortable position while eating.

One portable food tray per crewman is used to heat frozen food in large food cans, and to serve the complete meal (Fig. 124). Magnets, dispersed about the surface of the food tray, retain the reusable utensils while they are not in use. One utensil set consisting of knife, spoon, and fork is allocated to each crewman. Disinfectant-moistened pads, obtained from a galley-located tissue dispenser, will be used to cleanse the utensils after each use.

The wardroom food preparation and serving table also provides support in other activities, such as writing and playing games, when body restraint and restraint for objects are needed. A window in the wardroom, designed to accommodate several experiments which require exterior viewing, also affords a look to the ouside for the crew (Fig. 125).

The Waste Management Area (toilet) is shown in Fig. 126. Waste management facilities presented a unique challenge to spacecraft designers. In addition to collection of liquid and solid human wastes, there is a medical requirement to dry all solid human waste products and to return the residue to Earth for examination. Liquid human waste (urine) will be sampled and frozen for return to Earth. Total quantities of each astronaut's liquid and solid wastes will be precisely measured.

FIGURE 123.—Hatch for disposal of waste in former oxygen tank.

In the case of solid waste, a bag with a special filter is installed in the suction line of the toilet, allowing gases to pass through, and retaining only feces. Upon collection, the weight will be determined on a special spring-pendulum scale, and the bag will be placed in an electrically heated compartment where the contents are dried. The bag containing solid residue will then be stored for return to Earth.

Liquid waste will be processed with a centrifugal device installed in the suction line which imparts to the liquid sufficient force to actuate a precise liquid quantity meter. A constant volume (approximately 120 milliliters) of the liquid will then be separated and stabilized by freezing until it is returned to Earth.

Cabin air will be drawn into the toilet and over the waste products to generate a flow of the waste in the desired direction. The air will then be filtered for odor control and for antiseptic purposes prior to being discharged back into the cabin.

The washing facility is also illustrated in Figure 127. The crewman will wet a washcloth by placing it over the water discharge, apply soap to the cloth, and bathe and rinse as he would at home. The wet cloth will be discarded after use. A body shower will be taken by each crewman about once a week, the limit being set by the water storage capacity on Skylab (six pints per shower). Water will be discharged through a showerhead at the end of a flexible hose. Showers will be taken inside a cylindrical compartment (Fig. 127). The floating water droplets will be driven into a water collection system by air flow. Towels and tissues will be supplied as well as antiseptic cleaning agents.

Electric or safety razors may be used for shaving. A mirror is provided above the lavatory (Fig. 126).

Each crewman is assigned a small space for sleeping, as shown in Figure 128. Because of the absence of gravity, sleeping comfort can be achieved in any position relative to the spacecraft; body support is not necessary. Sleeping, therefore, can be accommodated quite comfortably in a bag which holds the body at a given place in Skylab and also encloses it in a manner which is psychologically and physically pleasing.

Finally, the main floor includes an experiment work area in addition to the wardroom and the waste management and sleeping areas described above. This work area is shown in Figures 129 and 130. Medical experiments will be conducted in this area (Chapter V-3).

e. Electric Power System

Solar energy is the prime source of electric power on Skylab. Two systems of solar-electric cell arrays, one on the Workshop and one on the Apollo Telescope Mount, will be deployed after the Skylab cluster has reached orbit. The OWS system consists of two wing-like structures which are folded and packed against the Workshop wall during ascent. The ATM system, folded in a similar way during ascent, deploys in the shape of a four-bladed windmill (Fig. 131).

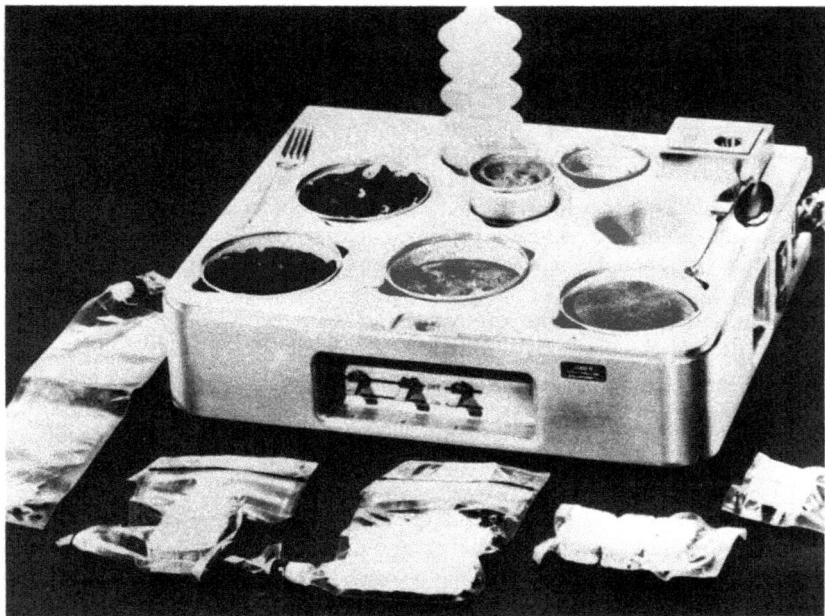

FIGURE 124.—Food tray and eating utensils for Skylab crew. By contrast, Apollo astronauts had only the food packets shown in the foreground.

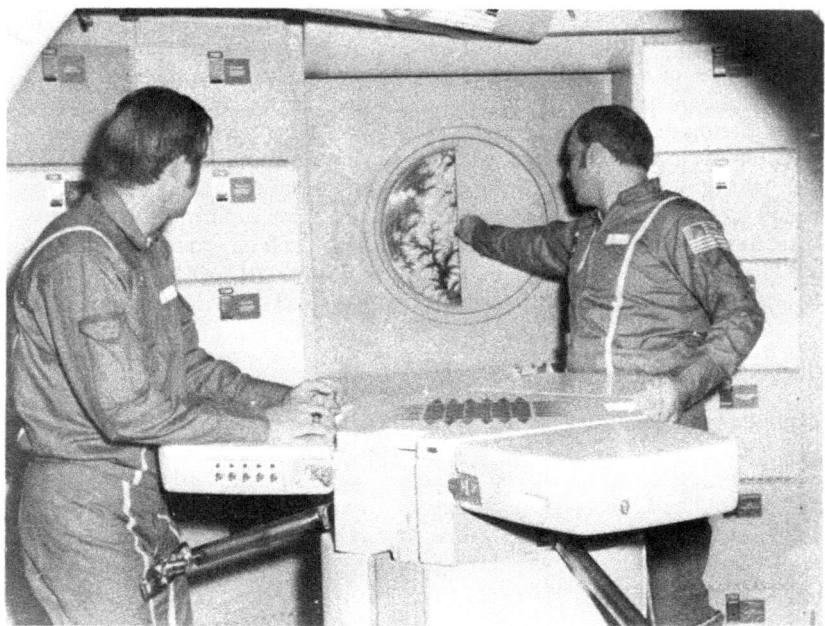

FIGURE 125.—Food table with leg restraints, and window to the outside.

Each system provides about 110 m² (1180 ft²) of active silicon cell area. Under ideal conditions of solar irradiation, each array will produce almost 12 kw of electric power. During the dark portion of each orbit, solar energy is not received, and storage batteries must provide power. Battery chargers, voltage regulators, power conditioning units, and distribution

FIGURE 126.—Waste management (toilet) and sample storage.

components will consume a certain portion of power. Under the influence of all these factors, the OWS array will produce an average usable power of about 3.8 kw, and the ATM array an average usable power of about 3.7 kw.

The Airlock Module will serve as the power center for the Workshop, the MDA, and the AM. Power from the OWS array is routed to the AM where batteries and power conditioning components are located. The ATM power conditioning equipment is mounted on the ATM rack; it will be controlled from the ATM Control and Display Console located in

FIGURE 127.—Shower compartment within Workshop.

110

the MDA. Although the two power systems are self-contained, they are interconnected in a parallel mode to allow maximum utilization of the available capacity. In addition to the permanently wired power-consuming systems, there are 28-volt utility outlets in the OWS, the AM, and the MDA for lights, tools, equipment, and even a vacuum cleaner.

Power for the Command and Service Module will be provided by hydrogen-oxygen fuel cells during ascent. After docking, the CSM power line will be connected to the Workshop network. On the return flight to Earth, the CSM will draw its electric power from batteries.

FIGURE 128.—Sleep compartment with sleeping bag.

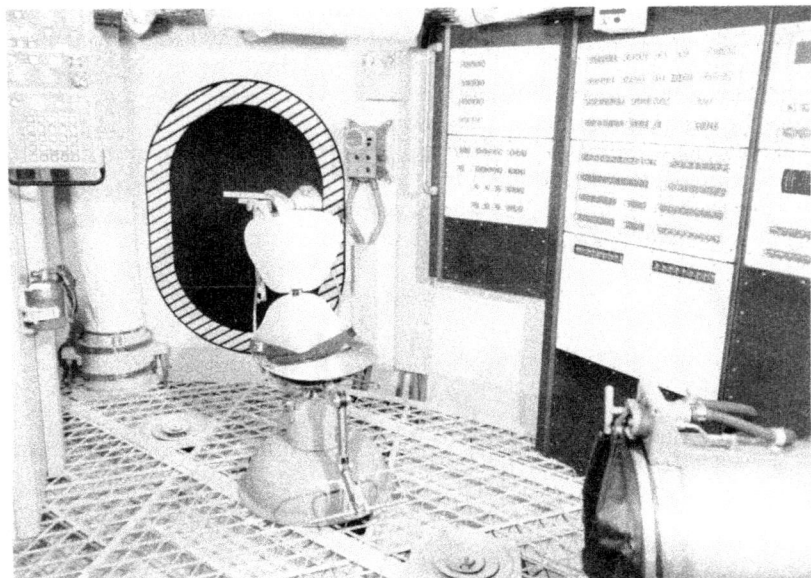

FIGURE 129.—Work area with litter chair for rotation of an astronaut (mockup).

FIGURE 130.—Work area with medical experiments, including Lower Body Negative Pressure system and wheelless bicycle (Ergometer).

FIGURE 131.—Unfolding of solar arrays of Apollo Telescope Mount (ATM) in orbit.

CHAPTER V

Research Programs on Skylab

Skylab, combining the unique features of an orbiting spacecraft with the convenience of a roomy, well-equipped laboratory, offers an unprecedented opportunity for research in a number of scientific and technical areas. Almost 300 separate investigations, taking advantage of this opportunity, will advance knowledge in four major areas: science, Earth observations, zero-gravity technology, and the reaction of man to the environment of space.

In the space science area, major emphasis rests on solar observations, most of them with the instruments mounted on the Apollo Telescope Mount. The ATM will observe the Sun with eight instrments: the white light coronograph, two X-ray telescopes, three ultraviolet spectrographs, and two heliographs imaging the Sun in the red light of the H-alpha line. (The H-alpha line is the red line in the Balmer series of the hydrogen spectrum, 656.28 nanometers or 6562.8 Ångstrom units).[1] An ultraviolet and X-ray spectrograph will observe the Sun through a scientific airlock in the wall of the Orbital Workshop.

Several instruments will record the ultraviolet and X-ray emissions of stellar objects within our Milky Way galaxy.

Cosmic ray particles will be recorded, and the minute impact craters of micrometeoroids on polished metal plates will be studied after the exposed plates have been brought back to Earth.

Medical doctors and biophysicists will utilize the state of weightlessness on board Skylab for scientific investigations. The effects of complete absence of gravitational forces upon metabolism, growth, and division of cells, upon tissues and organs, upon development cycles, and upon the wake-and-sleep rhythms of animals, will be studied in several experiments.

Observations of the Earth's surface from orbit, one of the major objectives of Skylab, will be carried out by the EREP assembly (Earth Resources Experiment Package). It contains six different instruments which will view targets on Earth in visible light, in infrared, and with microwaves. These observations will cover large areas of the Earth in a very short time under identical lighting conditions; they will provide information on such large-scale variables as cloud cover, snow and water conditions, ocean state, crop conditions, vegetation growth, development of urban and rural areas, water pollution, land use, and other factors which are of vital importance in the interaction of man with his environment.

The effects of gravity, ever-present on Earth, are not observable in the environment of an orbiting spacecraft. Processes such as convection, mixing of dissimilar components, diffusion in fluids, heat conduction, flow patterns,

---

[1] (1 nanometer=1 nm=10^{-9} meter=10 Ångstrom units)

liquid surface forming, crystal growing, casting of composites, welding, and flame propagation, which are influenced by gravitational forces on Earth, will be different in space. Some of the familiar methods of manufacturing and assembly will require new techniques under space conditions; on the other hand, some processes which cannot be achieved on Earth, such as the alloying of metals with greatly different densities or the formation of certain glasses, may become easy under weightlessness. A series of experiments to study such processes will be carried out on Skylab.

For the firt time in the space program, Skylab will offer an opportunity to systematically study the problems of life and work of man under prolonged exposure to space conditions. Numerous experiments were prepared to observe physical and mental functions of the astronauts, environmental conditions inside and outside the spacecraft, habitability features of the Workshop, the utility of tools, interfaces between astronauts and instrumentation, and the functioning of auxiliary systems. Many of these experiments are biomedical in character; their results will help us understand how man will adapt to the unique environment of a laboratory in space and how future space stations and deep space probes should be equipped to assure a comfortable and productive existence for astronauts. At the same time, experiments in this program will teach us how to build and equip spacecraft of the future in such a way that they offer optimum technical conditions for scientific research, for Earth observations, for zero-gravity technology, and as a habitat for the astronauts.

Experiments on Skylab will be described in four groups according to their objectives: science, Earth observations, life sciences, and space technology. Sections on the Skylab student project and on the postflight evaluation of Skylab data will follow.

The numbers listed with each experiment are the official designations for the Experiment Program; see also Chapter VIII, Listing of Skylab Experiments.

1. SPACE SCIENCE PROJECTS

As an observing station in orbit, Skylab has attracted the interest of astronomers, physicists and biologists from the time it was first conceived. In fact, a number of crucial observations in these three areas of scientific endeavor are planned, with the expectation that results will greatly advance our knowledge and will lay the groundwork for further research. The most prominent package of scientific instruments on Skylab, the Apollo Telescope Mount, will permit a study of the Sun. Some observations are planned of stellar objects, and some studies will be made of phenomena near the Earth which are difficult or impossible to observe from the ground. Biological studies on the effects of weightlessness will be described under Life Science Projects, Chapter V–3.

a. Solar Studies

Skylab's total man-attended time of 140 days, spread out over a period of eight months, will provide a unique opportunity to observe the Sun and its many surface phenomena in wavelength regions which are not accessible from Earth. Fig. 132 shows a portion of the Sun as viewed from Earth in the red light of hydrogen. Skylab instruments will be able to view such active areas also in ultraviolet and X-ray light. In orbit, seeing conditions are always flawless; the image quality depends only on the resolving power of the optical system, the pointing stability of the instrument, and the capability of the sensor.

Eight different telescopes on ATM will observe details of the Sun in various wavelength regions, as indicated in Fig. 133 which shows the spectral coverage of the instruments and also the transmissivity of the Earth's atmosphere as a function of wavelength. ATM will attack a wide variety of problems in solar physics (Fig. 134) through coordinated observations with such instruments as a white light coronagraph which will photograph the corona out to about six solar radii, a spectrograph for the 97 to 394 nm range, a spectrometer-spectroheliometer for the 30 to 140 nm range, a spectroheliograph for the 15 to 62.5 nm range, two X-ray telescopes covering the 0.2 to 6 nm range, and two H-alpha cameras (656.3 nm) which will provide images of the Sun's disc in the red light emitted by excited hydrogen atoms (Fig. 135). All of these instruments are rigidly mounted on the spar inside the ATM canister (see Chapter IV–1–e). The canister can be fine-pointed within pitch and yaw movements of $\pm\,2°$ relative to the rest of the Skylab. These movements will permit the exact orientation of the telescopes to any point on the Sun; the solar disc subtends an angle of 0.5° when seen from points on or near the Earth.

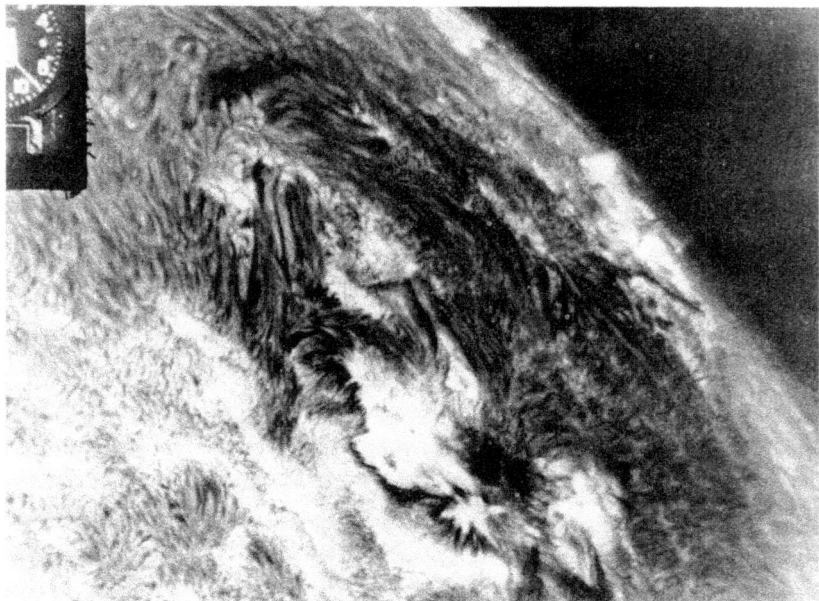

FIGURE 132.—Photograph of a portion of the sun, taken in the red light of the hydrogen alpha spectral line (625.3 nanometer or 6253 A). Courtesy of California Institute of Technology Observatory at Big Bear Lake, California.

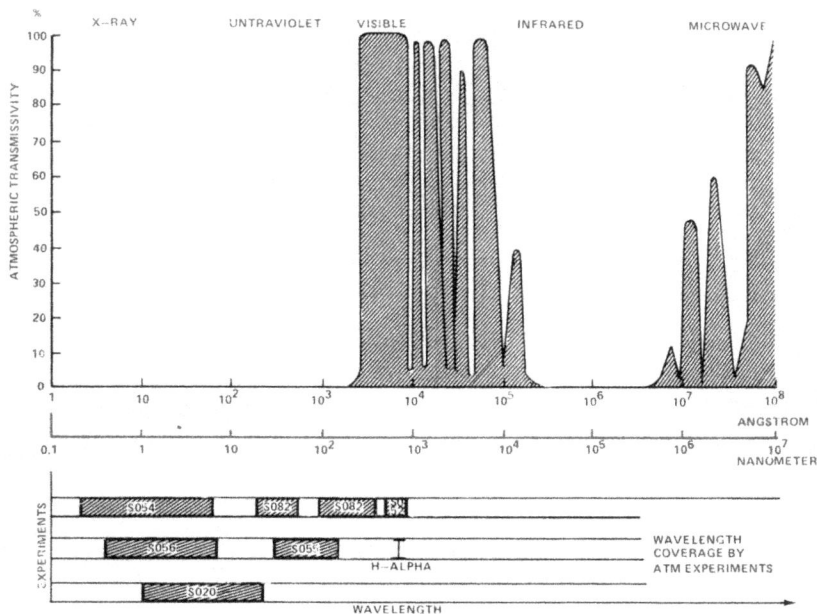

FIGURE 133.—Transmissivity of the atmosphere for various regions of the wavelength spectrum.

A single control and display console in the Multiple Docking Adapter adjacent to the ATM will permit manual operation and visual monitoring of all the experiments on ATM through selector switches, pointing controls, TV monitors, and a variety of indicators of experiment status, film usage, solar conditions, and other parameters (Fig. 99).

The role of the scientist astronaut on board Skylab will be to recognize and to point at targets of opportunity which promise a particularly high yield of scientific information; to survey, diagnose, report on, and possibly modify instrument performance; and to retrieve photographic film by extra-vehicular activties (EVA) for transport back to Earth.

Observing programs that will be carried out during the Skylab operation do not emphasize individual experiments, but broad problems in solar physics which will be attacked simultaneously by all investigators. Among these observing programs are the following (Fig. 136):

Chromospheric network and supergranulation.
Active regions; their morphology and development.
Solar flares.
Prominences and filaments.
Center-to-limb studies of the quiet Sun.
Observations of slowly varying phenomena over periods of days and weeks.

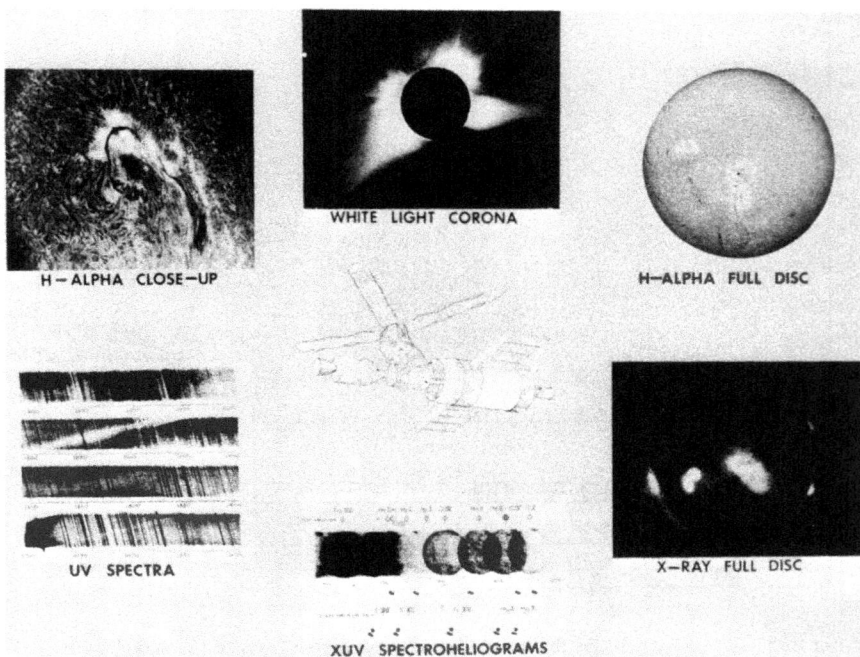

FIGURE 134.—Areas of solar research from Skylab.

Solar studies on Skylab will be complemented by a coordinated program of numerous ground-based observations, sounding rocket launches, and observations from other spacecraft.

During the Skylab mission, several sounding rockets will be used to launch subscale models of the ATM experiments S082 and S055 (ultraviolet spectrographs) for calibration purposes. The sounding rocket instruments will acquire data of a quiet solar region nearly simultaneously with the corresponding Skylab instruments. Since the rocket-launched payloads can be calibrated immediately before and after flight, the data obtained with these instruments will serve as reference for the near-simultaneous data acquired by the Skylab instruments. Resulting calibration factors will then be used in post-mission analysis of all Skylab data. Two calibration flights are scheduled for each of the two experiments during the Skylab flight; one during the first, and one during the second manned mission. If the first rocket flight should fail, a calibration flight will be scheduled during the third manned mission.

It is expected that the Skylab program of solar studies will bring a decisive increase in our knowledge of the Sun by extending our observing capability toward shorter wavelengths, toward higher resolution, and toward greater sensitivity. The fact that the same phenomena on the Sun will be observed

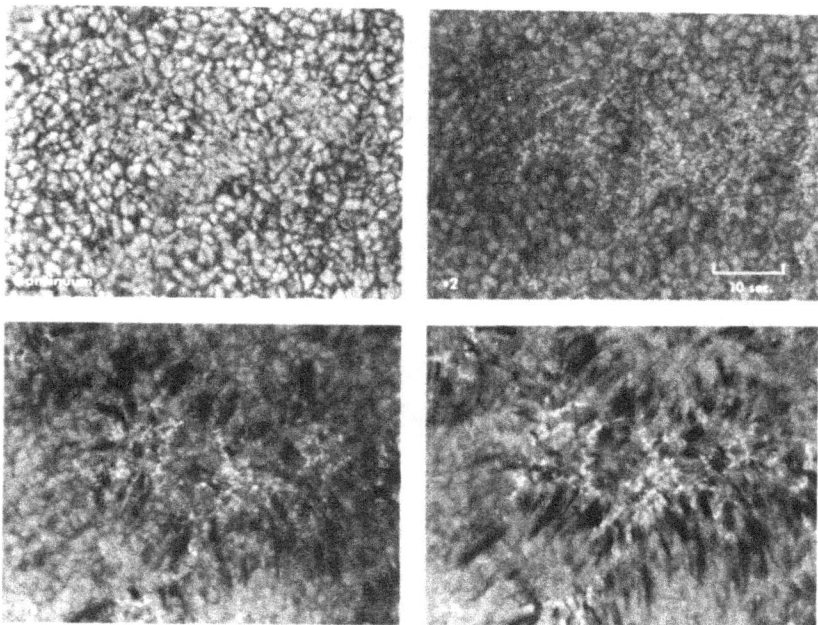

FIGURE 135.—View of a small area on the solar disc, taken in different wavelength regions. The granular structure on the top pictures is seen in white light; the bottom pictures in monochromatic (red) hydrogen light show areas of strong emission (light) and absorption (dark) of hydrogen clouds. Courtesy of Sacramento Peak Observatory, Air Force Cambridge Research Laboratories, Sunspot, New Mexico.

simultaneously by several different instruments on Skylab and from the Earth will greatly enhance the gain of scientific knowledge and understanding.

Observations of the Sun from Earth reveal three distinct outer layers on the Sun (Fig. 137). These are the very bright and almost opaque photosphere (about 5700° K temperature) with granules and sunspots; the more tenuous chromosphere (as low as 4000° K) with supergranules, plage areas, occasional violent outbursts in the form of solar flares, and other transient features such as vertical plasma jets called spicules; and the very tenuous corona which extends from the chromosphere (about 14,000 km—9,000 miles—above the photosphere) to a distance exceeding that of the Earth's orbit. Prominences, consisting of huge plasma clouds, represent extensions of the chromosphere far out into the corona. The inner corona has a temperature up to several million degrees K. The chromosphere is the region where the continuous spectrum radiated by the photosphere becomes the absorption spectrum with the black Fraunhofer lines.

Problem areas in solar physics which Skylab will attack and hopefully help clarify include such processes as the transport of energy from the photosphere through the chromosphere and into the super-hot corona; the origin,

500-721 O - 73 - 9

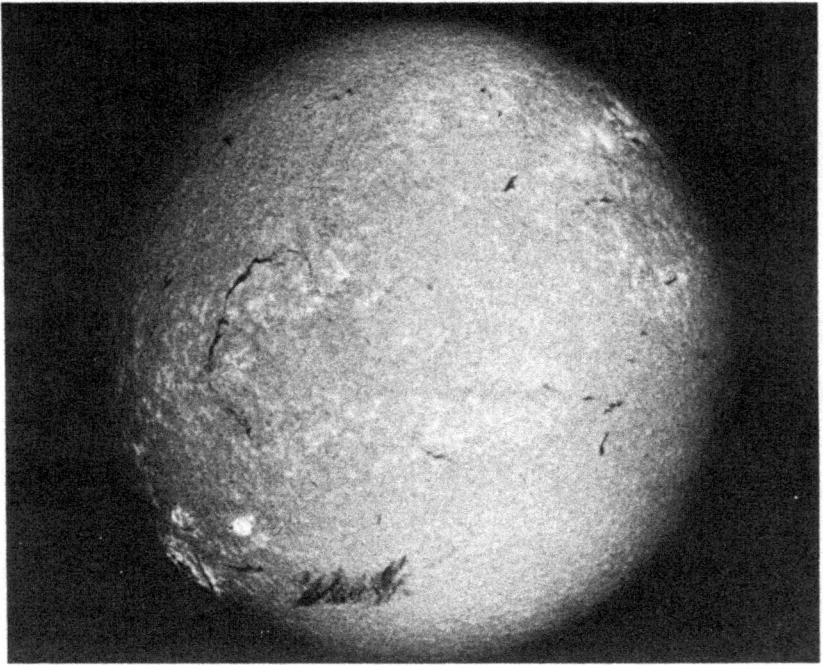

FIGURE 136.—Active areas on the Sun with violent outbursts of super-hot plasma, on July 26, 1972, photographed in the red light of the hydrogen-alpha line.

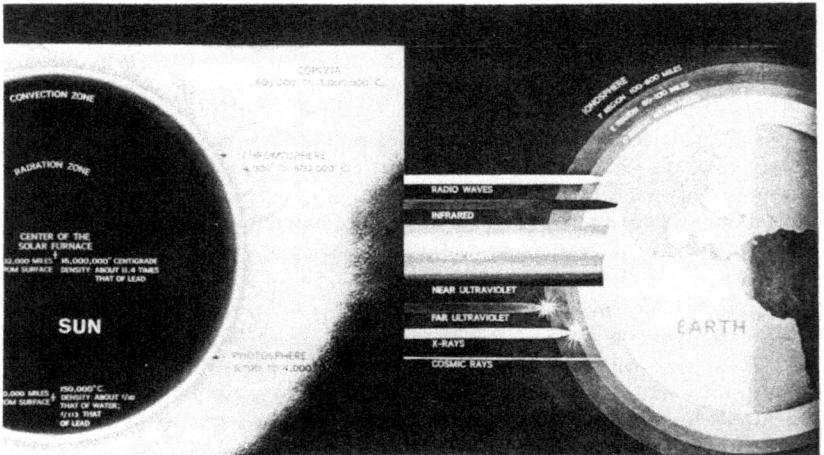

FIGURE 137.—Left: Major zones inside and outside the Sun; Right: absorption and transmission of various solar radiations in the Earth's atmosphere.

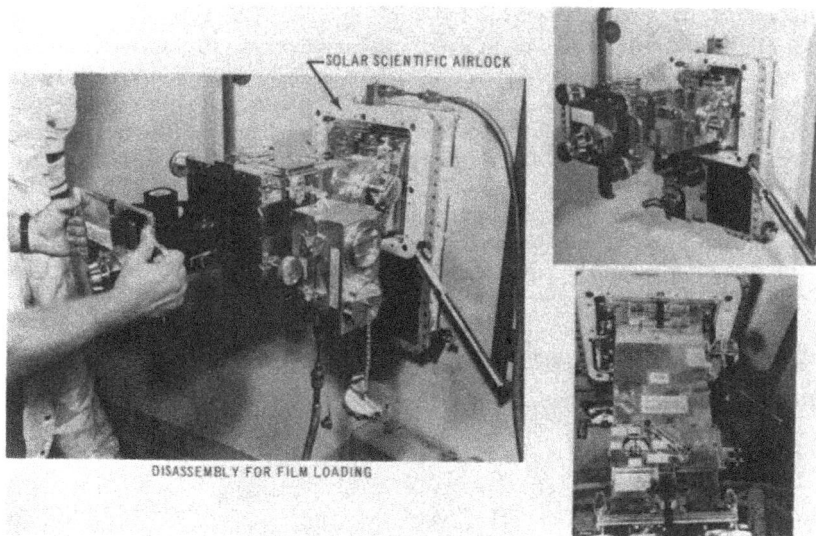

SOLAR SCIENTIFIC AIRLOCK

DISASSEMBLY FOR FILM LOADING

FIGURE 138.—Experiment S020, Solar photography with a spectrograph for soft X-ray and short ultraviolet radiation.

development, and energy-producing mechanism of solar flares; and the processes which produce the solar wind. Better knowledge of these solar processes will improve our understanding of the Sun's influence on our Earth environment, particularly on weather and climate. It will also expand our knowledge of plasma physics, a relatively young branch of science which will probably play a major role in future science and technology.

One of the solar instruments on Skylab (S020) is located in the Orbital Workshop; all the others are mounted on the central spar of the ATM canister. These experiments are described in the following sections.

S020, Ultraviolet and X-ray Solar Photography (Fig. 138)[1]

Principal Investigator:
Dr. Richard Tousey
Naval Research Laboratory
Washington, D.C.

Objective:

Record on photographic film a spectrum of the X-ray and ultraviolet radiation from the Sun in the one to 20 nanometer (10 to 200 Ångstrom) region, with modest angular resolution. Radiations in this spectral range are emitted by highly ionized atoms in the solar chromosphere and corona. They are indicative of high-temperature atomic and plasma processes which are extremely difficult to duplicate on Earth.

---

[1] Although this experiment is less significant than the ATM experiments, it is listed first because of the numbering system applied to the experiments.

Instrument:

Sunlight will enter a narrow slit and impinge upon a grating under a very small angle of incidence. Under conditions of grazing incidence, gratings reflect sufficient energy even in the one to 10 nm wavelength region (soft X-ray region) to make film recordings feasible when long-time exposures can be made. Thin metallic films in front of the slit will block out undesired ultraviolet and visible light.

The instrument is mounted in the solar airlock of the Orbital Workshop, facing the Sun. A finder telescope will enable the crew member to place the solar image on the slit of the spectrograph. Exposures will last up to one hour.

S052, White Light Coronagraph (Figs. 139, 140)

Principal Investigator:
Dr. R. MacQueen
High Altitude Observatory
Boulder, Colorado

Objective:

Obtain high resolution, high sensitivity photographs of the solar corona from 1.5 to 6 solar radii (300,000 km or 186,000 statute miles to almost three million km or 1.86 million statute miles above the solar surface). Study brightness, form, size, composition, polarization, and movements of the corona. Correlate the observations with solar surface events and with solar wind effects.

Instrument:

Coronagraphs are designed to block out the image of the Sun's disc and to take pictures of the faint corona which extends from the Sun far into space. Light scattering by optical elements and by structural surfaces must be carefully avoided. This instrument contains four coaxial occulting discs and photodetectors for alignment corrections. Pictures will be recorded on 35 mm film; they are taken either in unpolarized light or in one of three possible orientations of plane polarized light. Also, the instrument can operate in the "video mode" which will permit display for the astronauts or TV transmission to the ground.

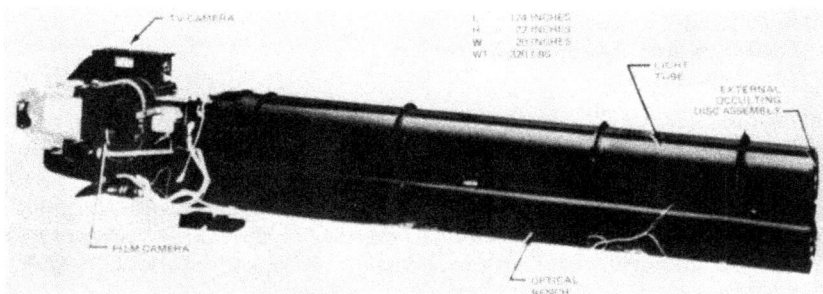

FIGURE 139.—Experiment S052, White Light Coronagraph observing the corona around the Sun.

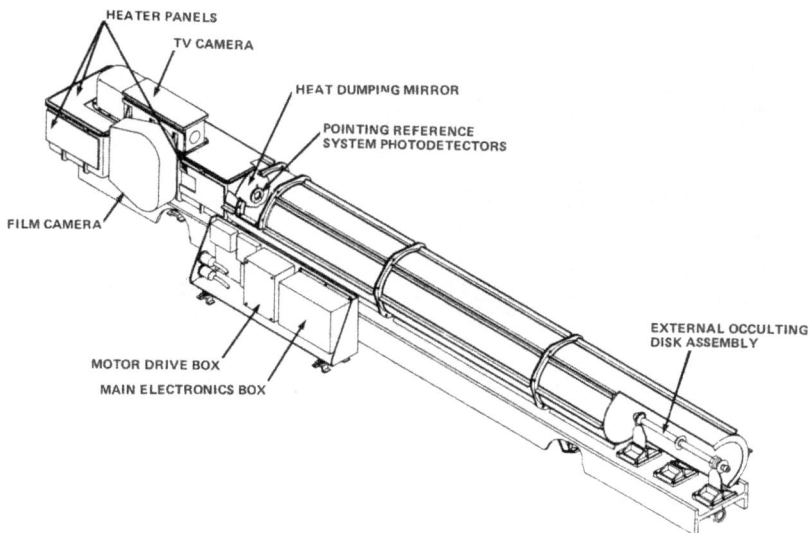

FIGURE 140.—Experiment S052, diagram showing constructional details.

Operation in four photographic modes is possible. In each mode, the shutter of the camera makes three exposures of 0.5, 1.5, and 4.5 seconds duration. In the first mode, this triple exposure is made at each of the four different positions of the polarization filter wheel. In the second mode, this same sequence of 12 exposures is repeated continuously for 16 minutes. In the third mode, the triple exposures repeat in fast sequence for 16 minutes, with the filter wheel in the "clear" position. The fourth mode will be the same as mode three, except that a shutter opening will occur every 32 seconds only; this mode will continue until manually stopped.

S054, X-ray Spectrographic Telescope (Figs. 141, 142)

Principal Investigator:
    Dr. Riccardo Giacconi
Acting:
    Dr. Giuseppe Vaiana
    American Science and Engineering Corporation
    Cambridge, Massachusetts

Objective:

Obtain X-ray images of the Sun over a wavelength range from 0.2 to 6 nm (2 to 60 Ångstrom). Record X-ray emissions of flares with a spatial resolution of two arc sec. Use selective filters and a transmission grating to obtain spectral information. Follow the evolution of active areas and correlate X-ray emissions with solar events observed in ultraviolet and visible light.

Solar X-rays are emitted from flares and also from other regions of activity, such as plage areas, prominences, and the corona. Two basic

Figure 141.—Experiment S054, Spectrographic telescope for X-rays.

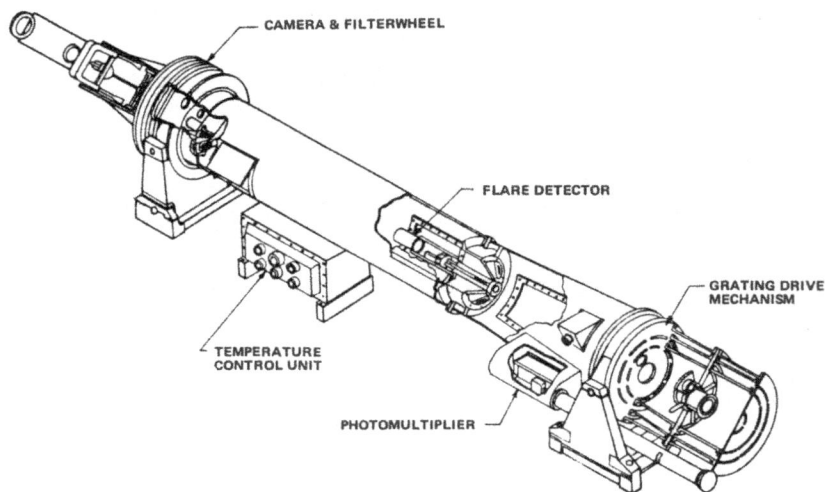

CAMERA & FILTERWHEEL

FLARE DETECTOR

GRATING DRIVE
MECHANISM

TEMPERATURE
CONTROL UNIT

PHOTOMULTIPLIER

Figure 142.—Experiment S054, diagram showing constructional details.

processes of the Sun seem to be responsible for most of the X-ray emission, the heating of plasmas, and the sudden acceleration or deceleration of electrons.

Instrument:

X-ray sources can be imaged with mirror optics utilizing very flat angles of incidence below about 0.5 degrees (Fig. 143). This experiment uses two cylindrical, coaxial mirrors of this kind with diameters of 31 and 23 cm (12.3 in and 9.2 in), with a total collecting area (two concentric rings) of 42 cm² (6.7 in²), and with a focal length of 213 cm (85 in). The transmission grating is mounted behind the rear end of the cylindrical mirrors; it will produce spectra on both sides of the zero-order image of a source (Fig. 144). A filter wheel mechanism will permit the insertion of selective filters into the path of the X-rays, thus providing broad-band spectral filtering of the flux. X-ray images will be recorded on 70 mm film.

A 7.6 cm (3 in) diameter, coaxial X-ray mirror will produce an X-ray image of the Sun on a scintillator crystal where it will be sensed by the

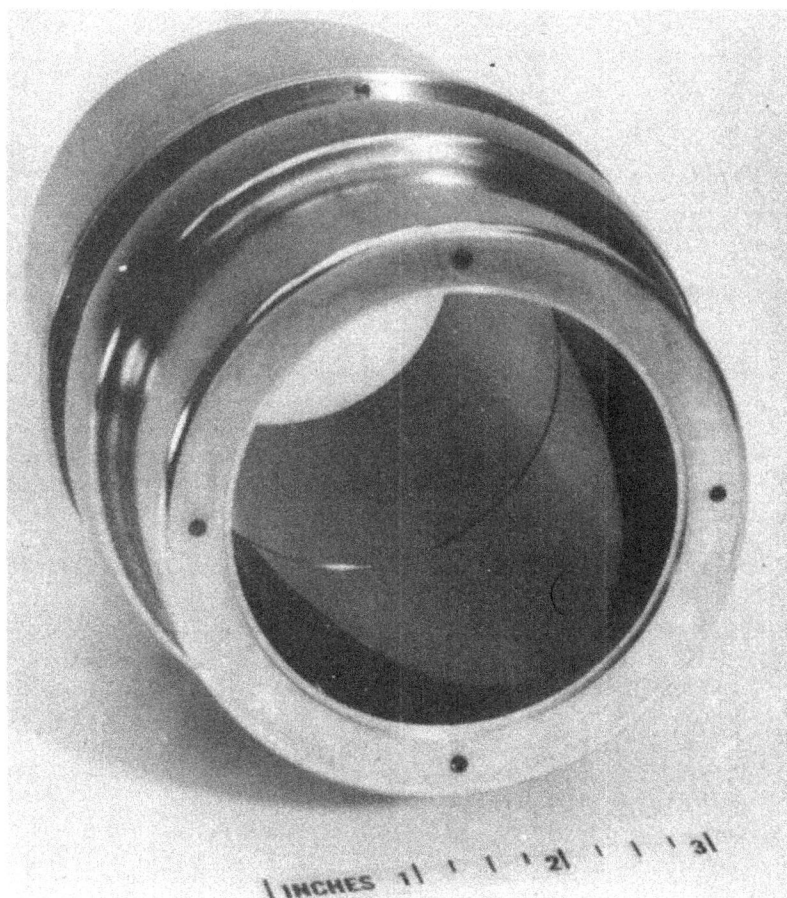

FIGURE 143.—Cylindrical Mirror for X-ray imaging telescope.

125

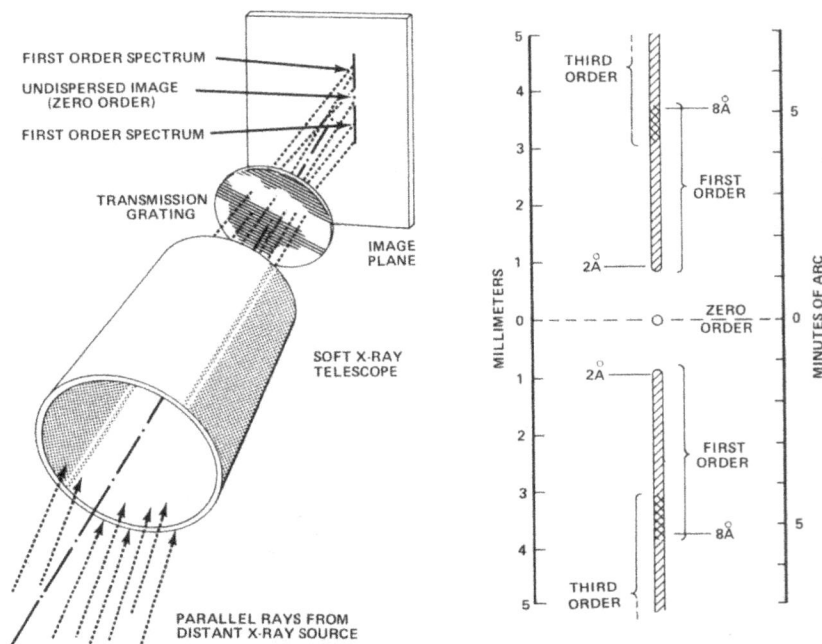

Figure 144.—X-ray telescope with grating, showing generation of spectra of individual X-ray sources.

photocathode of an image dissector tube. The output of this tube will be used for a visual display on the ATM console.

A photomultiplier tube, oriented toward the Sun, will measure the total solar X-ray flux; when a preset level is exceeded, an alarm for the astronauts will be given. The signal from this tube will also serve as a reference for the exposure setting of the film camera on the main telescope.

S055, Ultraviolet Scanning Polychromator/Spectroheliometer (Figs. 145, 146)

Principal Investigator:
Dr. Leo Goldberg, Director
Kitt Peak National Observatory
Acting:
Dr. Edward Reeves
Harvard College Observatory
Cambridge, Massachusetts

Objective:

Obtain photometric data of six spectral lines (O IV, Mg X, C III, O VI, H I, C II) [1] and the Lyman continuum [2] in the wavelength region from

---

[1] Three times ionized oxygen, nine times ionized magnesium, doubly ionized carbon, five times ionized oxygen, ionized hydrogen, and singly ionized carbon.

[2] Light below 912 Å emitted by hydrogen atoms.

FIGURE 145.—Experiment S055, Spectrometer for ultraviolet radiations.

30 to 140 nm (300 to 1400 Ångstrom) from 5 arc sec by 5 arc surface elements of the Sun. Also, obtain a spectral scan of the 30 to 140 nm region by tilting the grating.

Raster scanning of 5 arc min by 5 arc min areas will be achieved by rocking the primary mirror around two axes.

Radiations in this part of the spectrum are emitted by hot chromosphere and corona regions. The study of relative spectral line intensities will provide information about plasma composition, temperatures, and energy transfer processes in quiet and active solar phenomena.

Instruments:

An off-axis paraboloidal primary mirror will form a solar image on a 56 micron by 56 micron entrance slit of the spectrometer, corresponding to a 5 arc sec by 5 arc sec area on the Sun. Diffraction by a concave grating, ruled in gold with 1800 grooves per mm, will produce a spectrum on the Rowland circle [3] where seven photomultiplier detectors (Channeltrons) in fixed positions will simultaneously record the intensities of the six lines and the Lyman continuum. Bi-axial motion of the primary mirror will generate the desired raster scanning pattern (polychromator mode). In the grating scan mode, the primary mirror will remain fixed while the grating is tilted to scan the entire operating spectrum past one or more of the photomultiplier detectors. The signals from the detectors will be transmitted to the ground by telemetry.

---

[3] The Rowland circle determines the locations of slit, grating, and detector in a concave grating spectrograph.

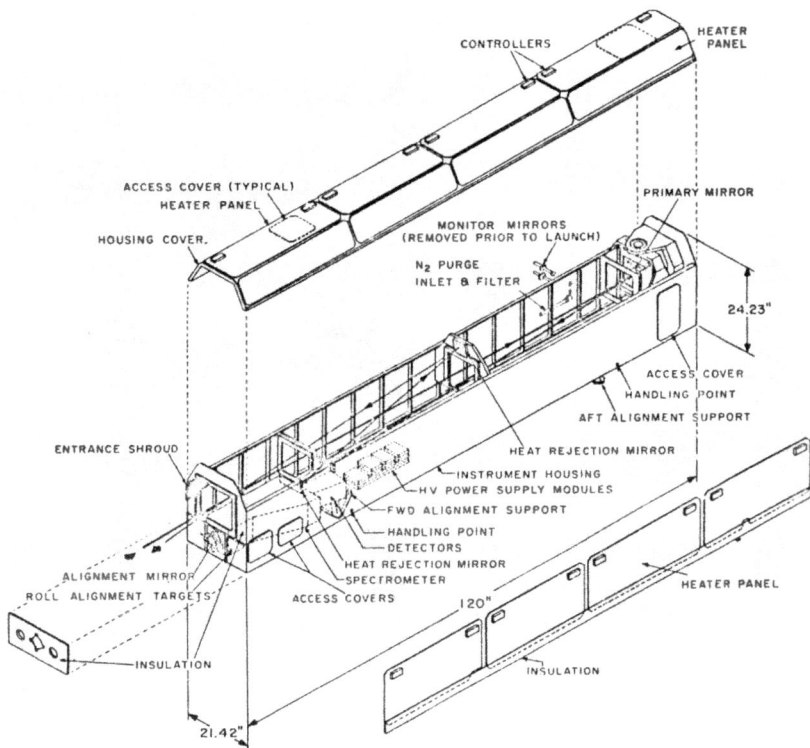

FIGURE 146.—Experiment S055, diagram showing constructional detail.

S056, X-ray Telescope (Figs. 147, 148)

Principal Investigator:

James Milligan
NASA-George C. Marshall Space Flight Center
Huntsville, Alabama

Project Scientist:

Dr. James Underwood
Aerospace Corporation
El Segundo, California

Objective:

Photograph the solar disc in X-ray light (0.6 to 3.3 nm or 6 to 33 Ångstrom) with high resolution in space and time, and modest spectral resolution. Attempt to obtain pictures during quiet and active periods. Monitor the total solar X-ray flux with proportional counters in the 0.2 to 0.8 nm (2 to 8 Ångstrom) and the 0.8 to 2.0 nm (8 to 20 Ångstrom) regions. Correlate the X-ray pictures with measurements of ultraviolet, visible, and microwave radiations from the Sun.

FIGURE 147.—Experiment S056, X-ray telescope.

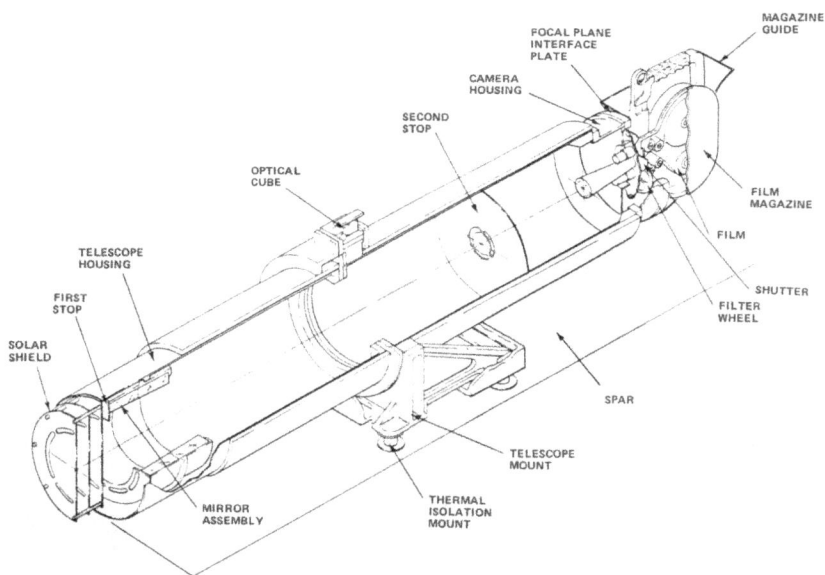

FIGURE 148.—Experiment S056, diagram showing constructional detail.

Observation of X-ray fluxes from the Sun will provide information on high-temperature regions and on interactions between hot plasmas and magnetic fields. Details of mass and energy transfer mechanisms and of the development of flares and prominences can be studied from X-ray measurements.

Instrument:

A cylindrical X-ray mirror with paraboloid-hyperboloid surface and grazing incidence, built of quartz, will form an image of the Sun on photographic film. Broad spectral discrimination will be achieved with five filters of beryllium, titanium, and aluminum, mounted on a filter wheel. Filters will be selected by astronaut decision.

Two proportional counters with mechanical collimators to improve the signal-to-noise ratio will continuously record the total X-ray intensity from the Sun in two wavelength regions. Their pulses will be pulse-height-analyzed and recorded on tape.

S082A, Extreme Ultraviolet (XUV) Spectroheliograph (Figs. 149, 150)

Principal Investigator:
Dr. Richard Tousey
Naval Research Laboratory
Washington, D.C.

Objective:

Record monochromatic images of the entire Sun in the emission lines of a spectral range from 15 to 62.5 nanometers (150 to 625 Ångstrom). Obtain information about composition, temperature, energy conversion and transfer, and plasma processes within the chromosphere and lower corona. Correlate these data with results from simultaneous observations in the other wavelength regions. Among the most intense lines in this extreme ultraviolet region are those of helium, oxygen, neon, magnesium, and iron.

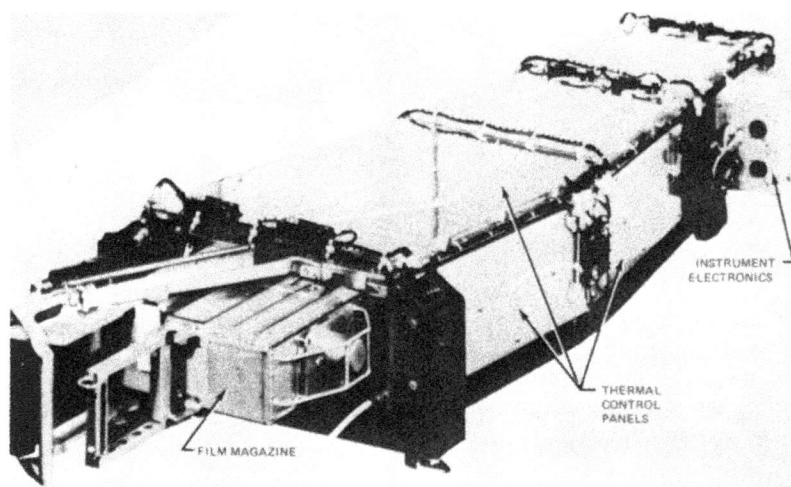

FIGURE 149.—Experiment S082A, Spectroheliograph for extreme ultraviolet radiations.

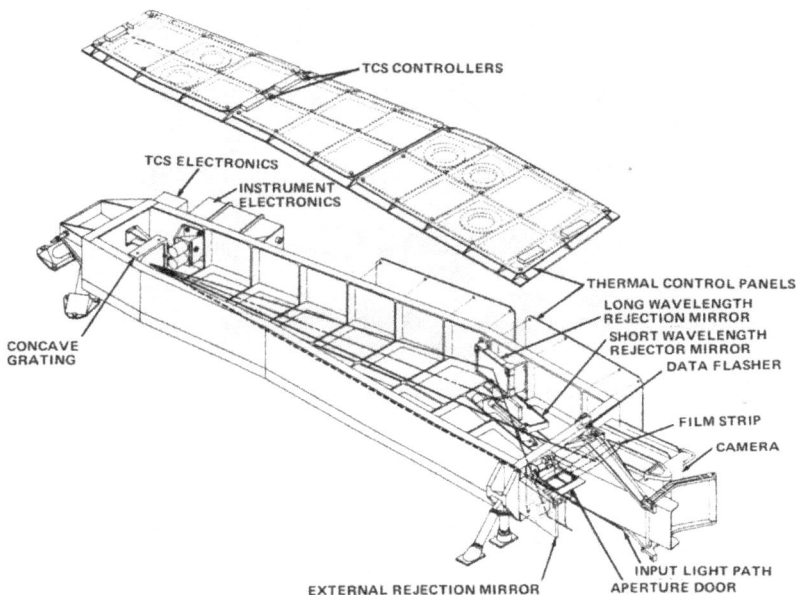

FIGURE 150.—Experiment S082A, diagram showing constructional detail.

Instrument:

Imaging of the Sun and generation of the spectrum will be achieved by a single concave mirror of 2 m (80 in) focal length, ruled in gold with 3600 lines per mm. Monochromatic, overlapping solar images of 18.6 mm (0.75 in) diameter will be formed on a film strip. Two spectral ranges of 15 to 35.5 nm and 32.1 to 62.5 nm (150 to 335 Å and 321 to 625 Å) will be photographed separately, with two angular positions of the grating. The unused part of the solar spectrum will be reflected out into space in order to avoid unnecessary heating of the instrument. A thin aluminum filter in front of the film will keep stray light out. Four film cameras, each loaded with 200 film strips, will be used; a crew member will exchange cameras by extravehicular activity (EVA).

S082B, Ultraviolet Spectrograph (Figs. 151, 152)

Principal Investigator:
    Dr. Richard Tousey
    Naval Research Laboratory
    Washington, D.C.

Objective:

Obtain UV spectra (97 to 394 nm or 970 to 3940 Ångstrom) of small portions of the solar surface with high spatial and spectral resolution. Photograph spectra at various locations on and off the disc and across the limb, from 12 arc sec below to 20 arc sec above the limb. Try to obtain spectra of flares and other active areas on the Sun.

FIGURE 151.—Experiment S082B, Spectrograph for ultraviolet radiations.

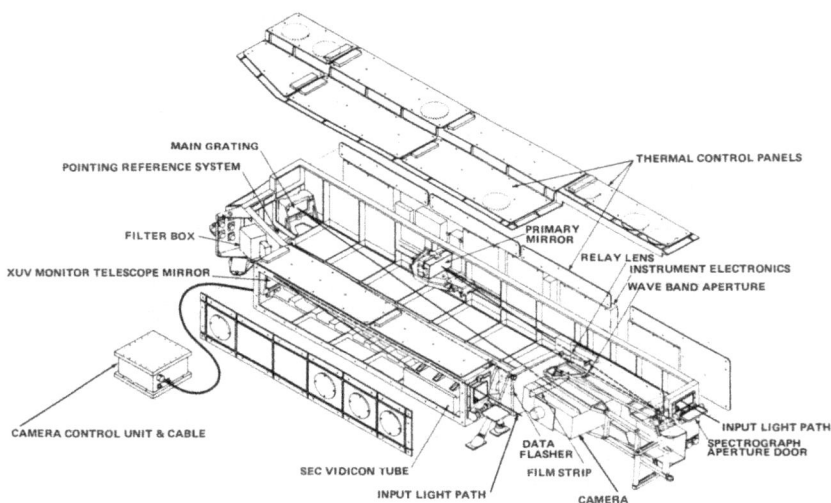

FIGURE 152.—Experiment S082B, diagram showing constructional detail.

Information on the change of the solar energy transportation mode from convection to plasma-dynamic shock waves will be derived from these observations. Also, details of structure, density, and temperature of the chromosphere and the lower corona will be studied.

Attached to this instrument will be the Extreme UV Monitor, providing a video image of the full solar disc in a broad spectral band (17 to 55 nm or 170 to 550Å), for coarse pointing and reference purposes.

An off-axis paraboloidal mirror will generate a solar image in the plane of the entrance slit of the spectrograph; the image on the reflecting slit jaws [1] will be viewed by a white-light TV system for pointing, selection, and verification by the astronauts.

---

[1] Se pg. 133, Footnote.

Instrument:

For reasons of stray light elimination, a predisperser grating assembly with two gratings will generate a light beam containing only the desired wavelength regions. The main grating, a concave mirror ruled at 600 grooves per mm, will produce a spectrum on photographic film with a resolution of 0.004 nm (0.04 Ångstrom) in the 97 to 197 nm (970 to 1970 Å) range and a resolution of 0.008 nm (0.08 Ångstrom) in the 194 to 394 nm (1940 to 3940 Å) range. The entrance slit will admit light from a 2 arc sec by 60 arc sec area on the Sun.

Several operational modes can be selected by the crew members, such as the boresight mode which will permit an astronaut to point at a specific area on the solar disc, the limb scanning mode which will produce a sequence of exposures across the limb by stepwise angular motion of the primary mirror, and the flare mode in which the instrument will take a preprogrammed series of exposures of flares or other active areas when commanded by a crew member. Pointing of the instrument to a desired area will be accomplished by moving the experiment canister with the hand controller on the ATM Control Console. Using the solar limb as a reference, the movable crosshair in one of the hydrogen-alpha telescopes will be adjusted so that its position coincides with the limb at the same time that the limb falls on the spectrometer slit, as viewed on the slit jaw image [1] with the white light TV system. After adjustment, the crosshair in the hydrogen-alpha telescope will always mark that point on the image of the solar disc which falls in the middle of the spectrograph slit.

Hydrogen-Alpha Telescopes (Figs. 153, 154)

Responsible for Development:
NASA-George C. Marshall Space Flight Center
Huntsville, Alabama

Objective:

Two telescopes imaging the Sun in the red light of the hydrogen-alpha line (Balmer series) will provide a visual aid to the astronauts and a photographic record of solar conditions during ATM observing periods. Each of the telescopes has a mechanically movable crosshair; that of Telescope I will be aligned with the boresight of Experiment S055, that of Telescope II with the boresight of Experiment S082B. Alignment will be accomplished by crew members, using the solar limb in two right-angled directions as reference system.

Hydrogen-alpha images of the solar disc will help crew members locate areas of solar activity and to recognize early stages of flare developments. Aiming the telescope and their crosshairs at such targets will automatically point those instruments which are aligned with the crosshairs to the same targets.

---

[1] The jaws of the slit of a spectrometer, being located in the focal plane of the imaging mirror of the spectrograph, carry an image of the light source, in this case the Sun. By making the surface of the slit jaws reflective, the solar image will be reflected and can be viewed by a TV camera. The spectrograph slit appears in this image as a thin black line. Inspection of the image will help identify the precise location on the solar disc where the spectrum was taken.

FIGURE 153.—Telescopic Camera for photographs in the red light of the hydrogen-alpha line.

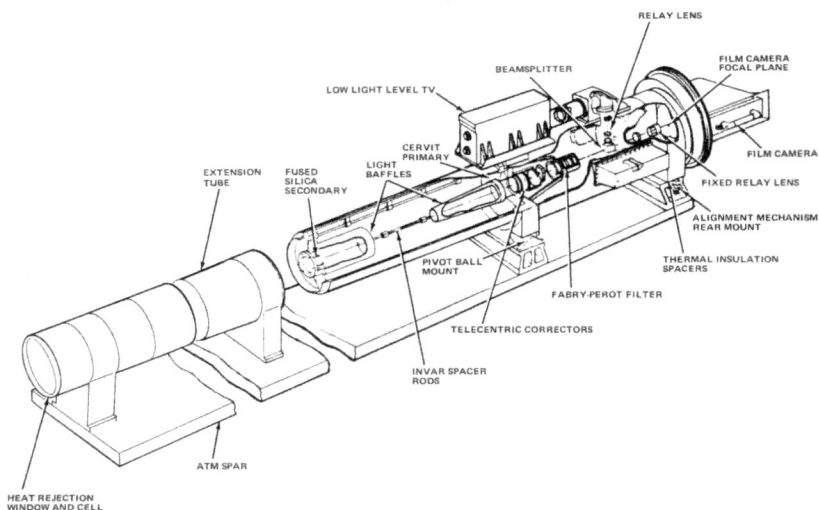

FIGURE 154.—Telescopic Camera for hydrogen-alpha photography, constructional detail.

Instruments:

Hydrogen-alpha Telescope I will provide simultaneous photographic and TV pictures; its resolution is one arc sec at a field of view of 4.5 arc minutes. Telescope II will operate only in the TV mode, with a resolution of about 3 arc sec. Each telescope has a zoom capability, varying the field of view between 4.5 and 15.8 arc min for telescope I and between 7.0 and 35 arc min for telescope II. Selection of the desired spectral line (656.28 nm or 6562.8 Ångstrom) is accomplished with a Fabry-Pérot filter which contains a solid glass flat with coated surfaces as interference gap. Bandpass is 0.07 nm (0.7 Å) for both telescopes. Polarizing elements in the optical path will permit polarization studies with both telescopes.

b. Studies in Stellar Astronomy

Although the total program in stellar astronomy on Skylab is modest, experiments in this program will pursue very interesting objectives. Two experiments will study ultraviolet low-dispersion spectra of star fields, nebulae, interstellar dust, and galaxies. Each photographic picture will contain the spectral images of numerous objects, permitting statistical evaluation of star populations. A third experiment, recording celestial X-rays with a large field-of-view instrument, will provide information on number and location of X-ray sources in various parts of the sky.

These experiments will provide important new data, leading toward astronomical observations with larger and more sophisticated telescopes on future space missions.

S019, UV Stellar Astronomy (Fig. 155)

Principal Investigator:
   Dr. Karl G. Henize
   NASA-Lyndon B. Johnson Space Center, Houston, Texas
      (formerly of Northwestern University, Evanston, Illinois)

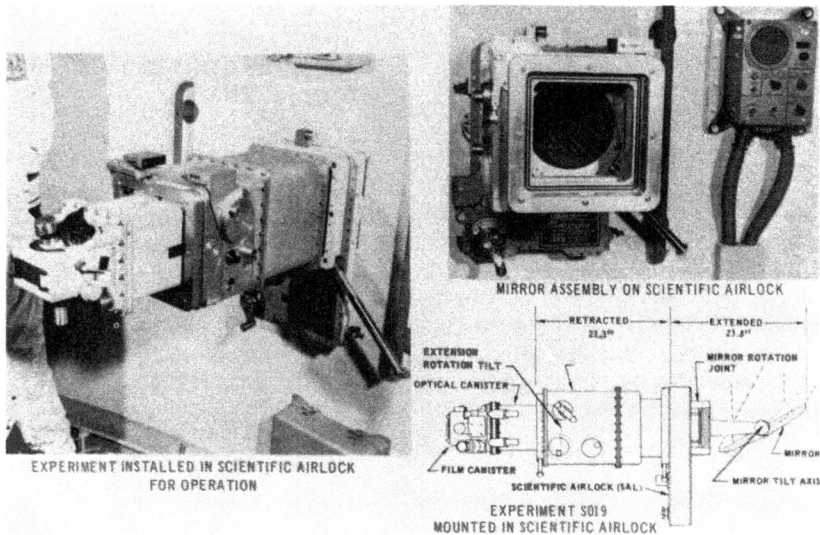

FIGURE 155.—Experiment S019, Camera with prism for ultraviolet star photography.

135

Objective:

Obtain ultraviolet spectra from stars, using a reflecting telescope and an objective prism in front of a 35 mm camera. The image of each star will be drawn out into a small spectrum. Evalute large numbers of spectra for spectral classes, temperatures, and compositions of stars. Obtain spectra of nebulae, interstellar dust, and stellar gas shells.

Instrument:

Mounted in the anti-solar airlock of the Orbital Workshop, the telescope will look at different portions of the sky by means of a movable flat mirror. Photographs will be taken only while Skylab is on the dark side of its orbit. The field of view of the system is so large that a number of star spectra are photographed with each exposure. Films will be developed and evaluated on the ground.

The telescope has a 15 cm (6 in) mirror, and a field of view of 4° by 5°. Several different prisms can be inserted in front of the telescope, depending on the desired spectral resolution and sensitivity. The instrument is sensitive in the spectral region from 300 to 140 nm (3000 to 1400 Å). Details can be resolved to about 20 arc seconds.

S150, Galactic X-ray Mapping (Fig. 156)

Principal Investigator:
Dr. William L. Kraushaar
University of Wisconsin
Madison, Wisconsin

Objective:

Survey selected portions of the sky for X-ray sources in the range from 0.2 keV (6 nm or 60 Å) to 10 keV (0.12 nm or 1.2 Å), and determine the location of sources with an accuracy of 20 arc minutes.

This experiment will provide knowledge of the existence and variability of celestial X-ray sources, and of mechanisms of the generation and absorption of X-rays in space.

Instrument:

The instrument package consists of nine gas-filled proportional counters, surrounded by 13 additional proportional counters operating as an active anti-coincidence shield. Mechanical collimators above the counters, admitting X-rays only within narrow angles, will determine the directions from which X-rays arrive. As the cluster changes its attitude, the instrument receives radiations from different areas of the celestial sphere. Star sensors will provide directional information, which is recorded on tape together with the X-ray data, and transmitted to Earth for analysis.

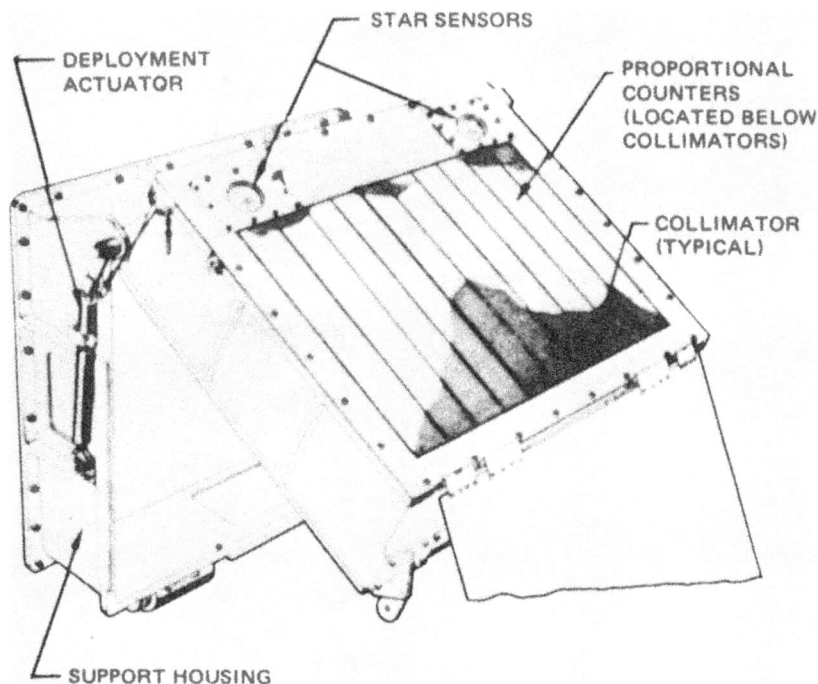

FIGURE 156.—Experiment S150, Detector for the recording of X-ray sources within our galaxy.

S183, Ultraviolet Panorama Telescope (Fig. 157)

Principal Investigator:
  Dr. Georges Courtès
  Laboratoire d'Astronomie Spatiale
  Marseilles, France

Objective:

Obtain UV photographs with spectral information of selected stars, star fields, stellar clouds, and galaxies. This experiment will furnish a large amount of data not avialable from Earth. Star fields in two UV bands (150 to 210 nm and 270 to 330 nm or 1500 to 2100 Ångstrom and 2700 to 3300 Ångstrom) will show the distribution of stars and of other celestial objects with strong ultraviolet emission. Color indices of these objects will be calculated from the photographs, and interstellar reddening will be determined from the color indices for about 1000 stars. Color indices will also be determined for star clusters, unresolved areas of the Milky Way, and selected galactic nuclei; statistical interpretation of large stellar populations will thus be possible.

The UV Panorama Telescope will be mounted in the antisolar scientific airlock for operation, similar to Experiment S019. Scanning of the sky will be achieved with the articulated mirror used for S019.

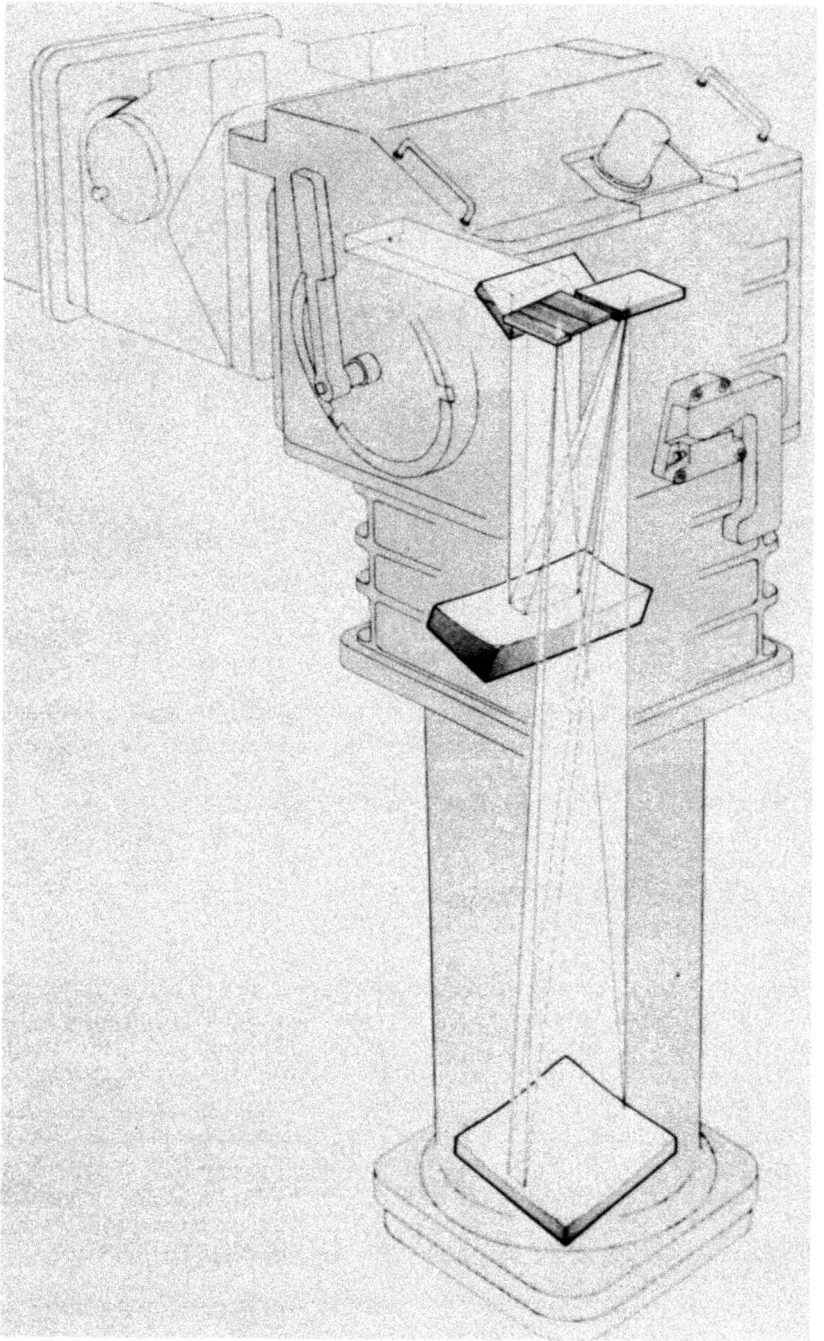

FIGURE 157.—Experiment S183, Telescopic camera for ultraviolet star photography.

138

Instrument:

Spectral photometry of star fields requires the combination of a wide angle imaging system with a spectrum-producing element. The UV Panorama Telescope uses reflecting mirrors and a plane grating. In order to make the system insensitive against angular drifting of the pointing direction, a mosaic of small lenses is placed in the focal plane of the telescope. A photographic plate in the focal plane of these small lenses will record minute images of the entrance pupil of the telescope. Only those of the small lenses which receive a star image in the focal plane of the telescope will produce bright pupil images on the plate.

As shown in Fig. 158, the grating will disperse the beam of light in such a way that the spherical mirror receives a spectrum. Two rectangular openings in a diaphragm in front of this mirror will admit only the two desired wavelength regions, with the result that each star produces two images, one for each wavelength region.

c. Space Physics Studies

Several physical phenomena in near-Earth space will be studied by Skylab experiments which will profit from the relatively long stay-time in orbit, the large weight-carrying capability of the spacecraft, and the presence of astronauts.

Cosmic ray particles, for which the atmosphere represents an almost impenetrable barrier, are numerous in empty space. Layers of photographic emulsion on Skylab, exposed to cosmic radiation, will trap such particles. Their tracks in the emulsion can be made visible by photographic development.

Faint luminosities in interplanetary space, caused by the scattering of sunlight from finely dispersed dust grains, have been observed from Earth. However, light from the Sun and other celestial objects, and also light from Earthbound sources, is scattered by the atmosphere to form a background of sky

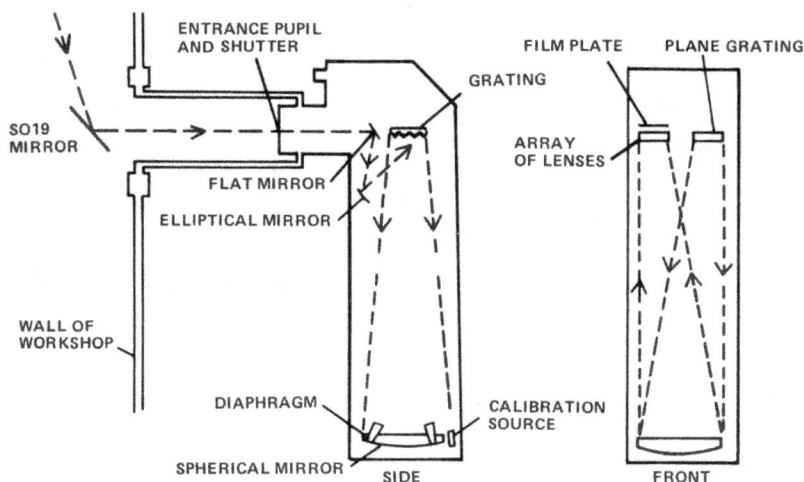

FIGURE 158.—Experiment S183, diagram with constructional detail.

brightness even at night which makes the observation of faint luminosities almost impossible. Observations from Skylab will avoid these difficulties.

Most of the meteoroids entering the Earth's atmosphere burn up before they reach the surface. However, in free space even small meteoroidal particles present a potential hazard to spacecraft because they can inflict damage to surfaces and container walls. Skylab will carry an experiment that will record impacts of micrometeoroids. It is expected that exposure times offered by Skylab will be long enough to permit statistical evaluations of abundance and mass of at least small micrometeoroids in the submicrogram category.

S009, Nuclear Emulsion Package (Fig. 159)

Principal Investigator:
Dr. Maurice M. Shapiro
Naval Research Laboratory
Washington, D.C.

Objective:

Record the tracks of cosmic ray particles as the particles penetrate a stack of photographic emulsion layers consisting of gelatin and silver bromide. Study the relative abundance of particles as a function of their masses.

Cosmic ray particles are fast moving nuclei of chemical elements, probably originating in thermonuclear reactions of certain stars. Of particular interest will be the tracks of heavy nuclei. Such particles do not normally penetrate the Earth's atmosphere because they quickly lose their energy in ionization and in strong interactions with other nuclei.

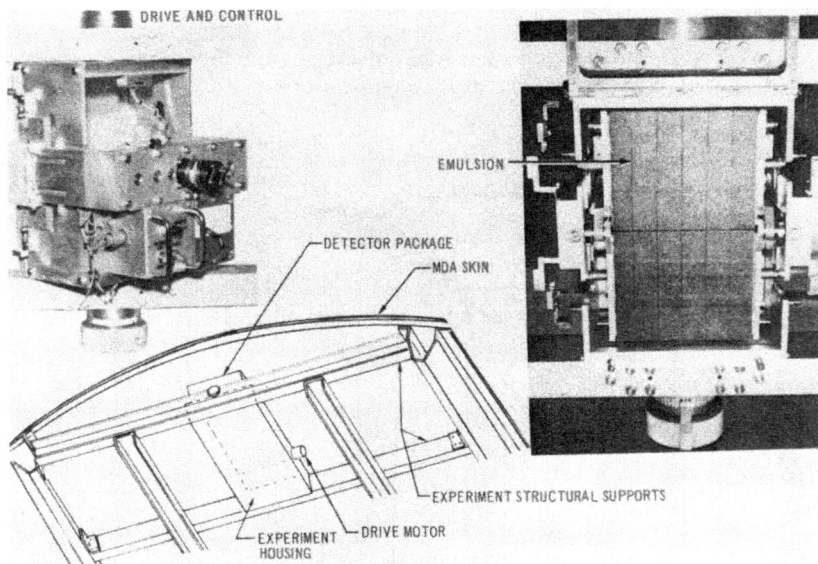

FIGURE 159.—Experiment S009, Emulsion layers for the detection of cosmic ray particles.

Two stacks of emulsion layers are arranged like the two sides of an open book. After arrival in orbit, one of the astronauts will open the "book" and deploy it inside the Multiple Docking Adapter, pointing toward outer space. While Skylab is moving through areas of high background radiation, the "book" will be closed. The emulsion layers will be exposed for about 240 hours during the first manned period and returned to Earth with the first crew.

Upon photographic development, the tracks of particles in the emulsion turn black because nuclear particles activate silver bromide crystals in their paths, similar to the way light activates an ordinary photographic emulsion. The thickness of a nuclear track corresponds to the ionization rate of the particle, which in turn is a function of its charge-to-mass ratio and of its energy.

S063, Ultraviolet Airglow Horizon Photography (Fig. 160)

Principal Investigator:
Dr. Donald M. Packer
Naval Research Laboratory
Washington, D.C.

Objective:

Photograph the Earth's ozone layers and the horizon airglow in visible and ultraviolet light. Take pictures in reflected sunlight and at night.

These observations will provide information on oxygen, nitrogen, and ozone layers in the Earth's atmosphere, and on their variations during night-and-day cycles.

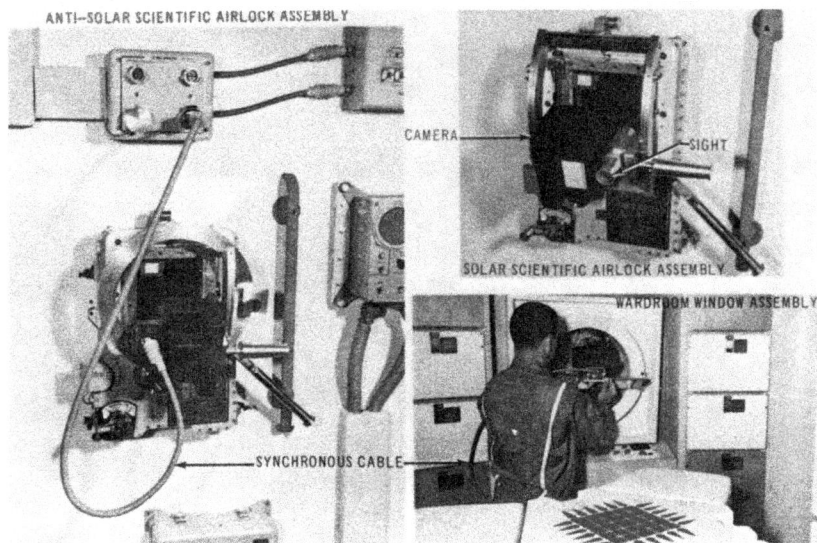

FIGURE 160.—Experiment S063, Camera to photograph the air glow in the Earth's atmosphere.

Instrument:

Pictures will be taken with 35 mm cameras from three positions within the Orbital Workshop, the solar scientific airlock, the anti-solar scientific airlock, and the wardroom window. Ultraviolet pictures and visible light pictures will be taken simultaneously and, after return to Earth, correlated for detailed evaluation.

S073, Gegenschein and Zodiacal Light (Fig. 161)

Principal Investigator:
Dr. Jerry L. Weinberg
Dudley Observatory
Albany, New York

Objective:

Measure the brightness and the polarization of the night glow of the sky over a large portion of the celestial sphere in visible light. Determine the extent and the nature of the spacecraft's corona during daylight.

The night glow of the sky is caused by sunlight reflected from interplanetary dust which accumulates preferably in the plane of the ecliptic (Zodiacal Light). Opposite the Sun, this band of light widens into an elliptical spot about 10 degrees in diameter (Gegenschein). Ground observations of the night glow are severely hindered by the airglow layer in the atmosphere at an altitude of about 90 km (56 statute miles or 48

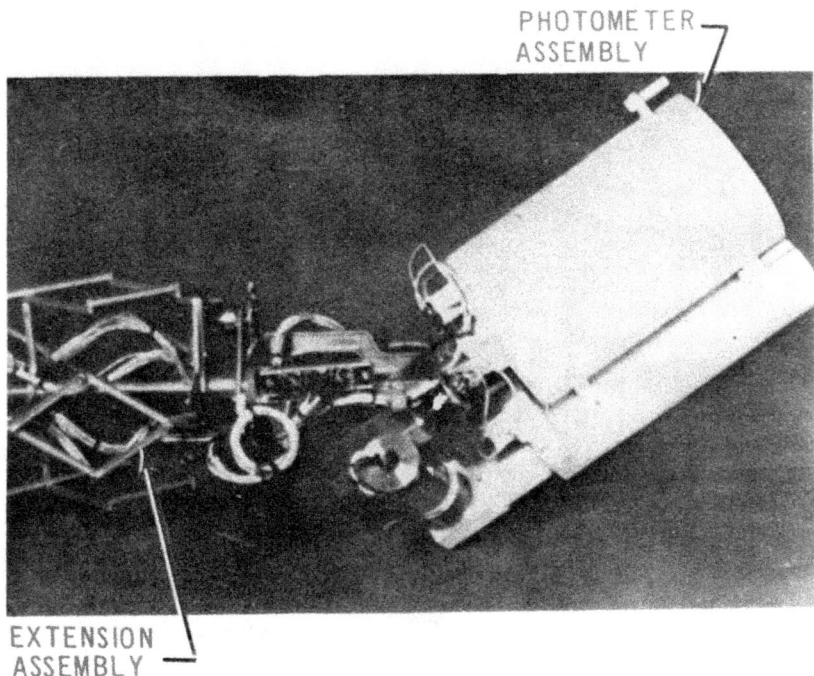

PHOTOMETER ASSEMBLY

EXTENSION ASSEMBLY

FIGURE 161.—Experiment S073, Camera to photograph faint luminosities in the sky.

nautical miles). Any halo around the spacecraft will curtail astronomical observations while Skylab is exposed to sunlight.

Instruments:

These observations will be made with the photometer built for the Contamination Measuring System T027 (Fig. 209). The photographic camera of T027 will provide star pictures for sky region identification. Observations will be made through both the solar and the anti-solar scientific airlocks.

S149, Micrometeoroid Particle Collection (Fig. 162)

Principal Investigator:
Dr. Curtis L. Hemenway
Dudley Observatory
Albany, New York

Objective:

Collect micrometeoroid particles on exposed surfaces and determine their abundance, mass distribution, composition, morphology, and erosive effects.

Instruments:

Four polished metal, glass, and plastic collector plates, each measuring 0.15 by 0.15 meters (6 inches by 6 inches) and mounted on a box-like support unit, will be exposed after deployment through one of the scientific airlocks on the universal extension mechanism of Experiment T027. Observations will be made on all three missions. After exposure times of a few days up to two months, the plates will be retracted and sealed in containers for protection. Analysis of the collected particles and of their effects will be performed on Earth after return of the plates in the Command Module.

Figure 162.—Experiment S149, Exposure of plates to collect micrometeoroidal particles.

S228, Trans-Uranic Cosmic Rays

Principal Investigator:
   Dr. P. Buford Price
   University of California
   Berkeley, California

Objective:

Record the tracks of heavy cosmic rays from iron (atomic number $Z=26$) to trans-uranic nuclei (atomic numbers $Z<92$) in layers of plastic material (Lexan). Determine the relative abundance of nuclei with atomic numbers above 26. Determine the energy spectrum of cosmic ray particles with atomic numbers from $Z=26$ to Z-numbers as high as possible. Particle energies from about 150 MeV [1] to more than 1500 MeV per nucleon are expected.

Cosmic ray particles with atomic numbers representing the heavy elements are extremely rare. However, measurement of their abundance and energy spectra will provide very valuable information about the synthesis of heavy elements in stars. Also, results of these measurements will be useful in the design of detectors for ultra-heavy cosmic rays proposed for future space projects.

Tracks of very heavy cosmic ray particles have been found in meteorites. Observations with high-altitude rockets have also provided some data on cosmic ray particles with high atomic numbers. The atmosphere prevents such particles from reaching the Earth's surface.

Instrument:

The heavy cosmic ray particle detectors are completely passive. They consist of stacks of identical layers of Lexan plastic sheets, mounted inside the thin walls of the Orbital Workshop. After returning to Earth, the Lexan sheets will be chemical etched. Etch pits develop at the top and bottom surfaces of each sheet where cosmic ray particles have entered and left the plastic layer. The length of a pit is proportional to the square of the ionization rate of the particle. By observing the track of a particle through many layers of a stack, atomic number and energy of the particle can be derived. The two stacks of Lexan sheets have a mass of 30 kg (66 lbs).

S230, Magnetospheric Particle Composition

Principal Investigators:
   Dr. Don L. Lind, Astronaut
   NASA-Lyndon B. Johnson Space Center
   Houston, Texas

   Dr. Johannes Geiss
   University of Bern, Switzerland

---

[1] MeV=million electron-volt, a measure of the energy of nuclear particles. (1 McV=1.602 x 10^{-13} Joule).

144

Objective:

Collections of helium, neon, and argon by exposing metal foils to the particle fluxes encountered by Skylab while traveling through the magnetosphere. In this sphere, which extends from about 160 km (100 statute miles or 90 nautical miles) to several Earth radii, the Earth's magnetic field strongly influences the trajectories of charged particles. The sources of charged particles to be collected by Skylab sensors include the Van Allen Belt radiation, possibly the solar wind, and the interstellar gas. Charged particles, particularly ions of atmospheric gases, may even reach orbital altitudes from the upper layers of the atmosphere.

Exposed foils will be returned to Earth after the second and the third Skylab missions. Implanted particles will be released by heating and analyzed with mass spectrometers. Particle energies can be estimated by employing layered foils, and by determining the depth of particle penetration by separate analysis of the different layers.

It is known that isotope ratios of noble gases in the Earth's atmosphere are very different from the ratios in the solar wind and probably in interstellar gas clouds. By determining isotope ratios in the collecting foils, particle fluxes of terrestrial, solar, and interstellar origin can be distinguished.

Instruments:

Charged particles of the solar wind have been successfully captured by metal foils in several Apollo experiments on the surface of the Moon and in rocket probe experiments. The experiment on Skylab will consist of sheets of collecting foils mounted on flexible plastic substrates. Collectors of aluminum, aluminum oxide, and platinum will be used. They will be mounted in the form of two "double cuffs" on a truss of the ATM supporting frame before launch. Each of the six collectors has a size of 0.35 by 0.48 m (14 by 19 inches).

After return to Earth, parts of each foil strip will be heated and finally melted in ultra-high vacuum systems, and the evolving gases will be analyzed in mass spectrometers.

Except for EVA retrieval, this experiment is entirely passive.

2. EARTH RESOURCES EXPERIMENT PROGRAM

Skylab offers an opportunity to expand remote sensing investigations of the Earth from orbit by utilizing relatively large aperture, flexible, high performance sensors, and crew members to operate the sensors. The Earth Resources Experiment Package (EREP) represents an experimental facility dedicated to that purpose.

Photography from spacecraft in Earth orbit in the visible and near-infrared wavelengths has proven valuable for mapping geographic and weather features over large areas of the Earth. Systematic application of remote sensing techniques using additional wavelengths may extend the usefulness of this capability to mapping of Earth resources and land uses. Resources subject to this type of study include crop and forestry cover, health of vegetation, types of soil, water storage in snow pack, surface or near-surface mineral deposits, sea surface temperature, the location of likely feeding areas for fish, etc. (Fig. 163).

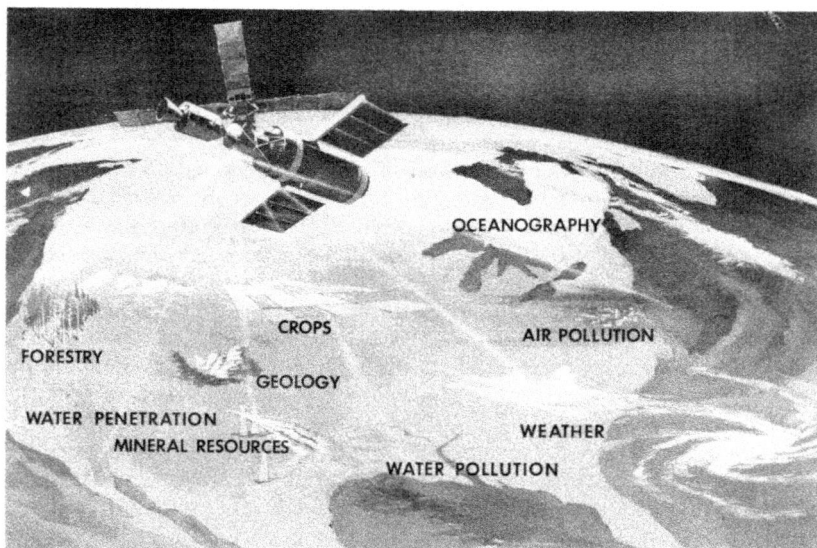

FIGURE 163.—Earth observation studies from Skylab.

In July 1972, the first Earth Resources Technology Satellite (ERTS–1) was launched into an approximately circular 1,000 km (600 mi.) polar orbit. ERTS–1 is providing, routinely, images of the U.S. and selected foreign areas in several bands encompassing the visible and infrared portions of the electromagnetic spectrum.

The Skylab Earth Resources Experiment Package is the next step in this important effort. EREP will use visible light and infrared photography. In addition, it will include imaging devices, electronic infrared spectography, and microwave radiometry surveys. EREP film and onboard storage techniques will permit large quantities of remotely sensed data to be gathered, and it will establish the feasibility and usefulness of Earth-survey techniques. The planned investigations will assess sensor types, designs, and capabilities needed to identify specific Earth resources and features. Requirements of future systems for specific applications can then be more firmly established. Methods for processing and interpreting data and the effects of atmospheric scattering and attenuation will also be defined.

Proposals to use EREP data have been submitted by university, government, and industry groups both within the U.S. and from abroad. From the more than 300 proposals which were submitted, 146 investigations were selected to comprise the EREP data-user program. These investigations have been subdivided into 170 individual tasks representing studies in 32 applications areas.

The EREP facility includes six sensors together with their associated support equipment. The support equipment includes the electronics to handle data from each sensor, a control and display panel, and primary and spare tape recorders.

EREP photography will greatly improve resolution of Earth resources phenomena by virtue of simultaneous exposures using six matched cameras, and by precise photometry that can provide more accurate knowledge of light intensity levels of various bandwidths in each survey photograph. The infrared spectroscopic survey will operate in wavelengths not recordable on photographic film and will provide data from which recognizable spectral signatures of the observed phenomena can be plotted. By simultaneously operating in frequencies transmitted by the atmosphere and in those attenuated by atmospheric moisture, atmospheric moisture density profiles can be generated.

The microwave radiometry equipment, because of its low sensitivity to atmospheric moisture, will provide an all-weather source of information on surface moisture and temperature and on vegetation distribution. Microwave radiometry over the oceans will provide information on wind and sea conditions.

Two crewmen are required to operate the sensors. The control and display panel contains individual switches which activate and select the operating modes for five of the EREP sensors. Also included are master power switches and the controls for the tape recorder. The assigned crewman will be responsible for operating each of the sensors from the control and display panel during its functional period. Other functions of the crew involve positioning the S190A boresighted camera array over the photographic viewing window, changing film/filter combinations, and changing lens dessicants and f-stops as required. Supply and return of film cassettes for the S190A array and the 16 mm camera, used with the S191 Infrared Spectrometer, will also be required. A crewman will operate slewing control of the tracking telescope to bring targets into the field of view of the S191 Spectrometer. The S190B Earth Terrain Camera will be deployed by the third crewman to view through the scientific airlock opposite the Sun in the Workshop. Film supply and retrieval to and from this instrument will also be accomplished by this crewman.

Another major function of the crew will involve coordination with ground-based activities and Mission Control to update EREP operations. Real-time decisions will be required because of local weather and cloud cover conditions. This important astronaut function will optimize the utilization of the EREP and, therefore, maximize Skylab's contribution of useful data to the Principal Investigators. An artist's conception of the ground coverage of the EREP is shown on Fig. 164.

The six sensors comprising the EREP package can be categorized under the headings of photography, infrared observations, and microwave studies. A description of each sensor follows.

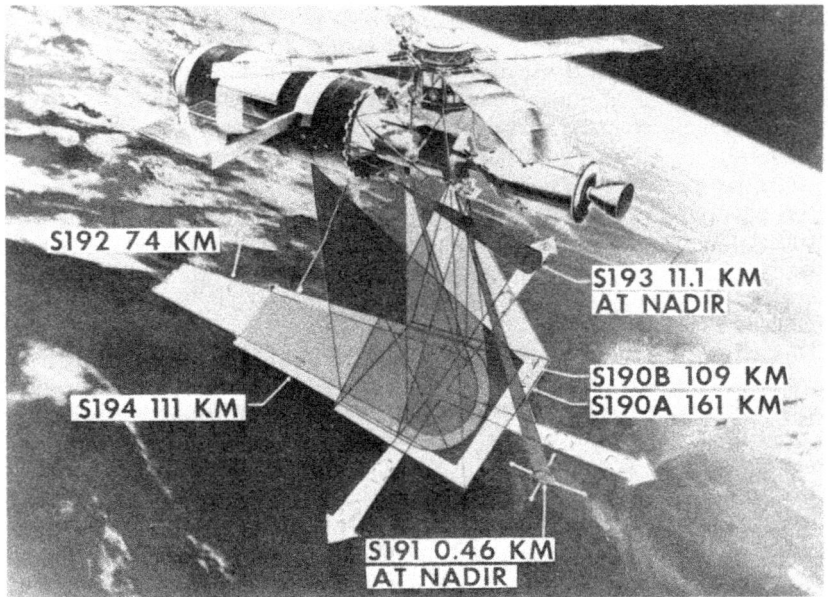

FIGURE 164.—Earth Resources Experiment (EREP) instrument coverage of surface areas from Skylab.

a. Photography

S190A Multispectral Photographic Cameras (Figs. 165, 166)

Project Scientist:
K. Demel
NASA-Lyndon B. Johnson Space Center
Houston, Texas

Objective and Instrumentation:

Obtain precision multispectral photography by selecting various 70 mm film/filter combinations for a wide range of Earth science studies. Detail not apparent in ordinary photography can be studied. Ground resolution will be 30 meters (100 ft); ground area coverage will be 161 km (100 statute miles) on a side. Observable features include water pollution, geological features, development of metropolitan and urban areas, and others. Six high-precision cameras with matched distortion and

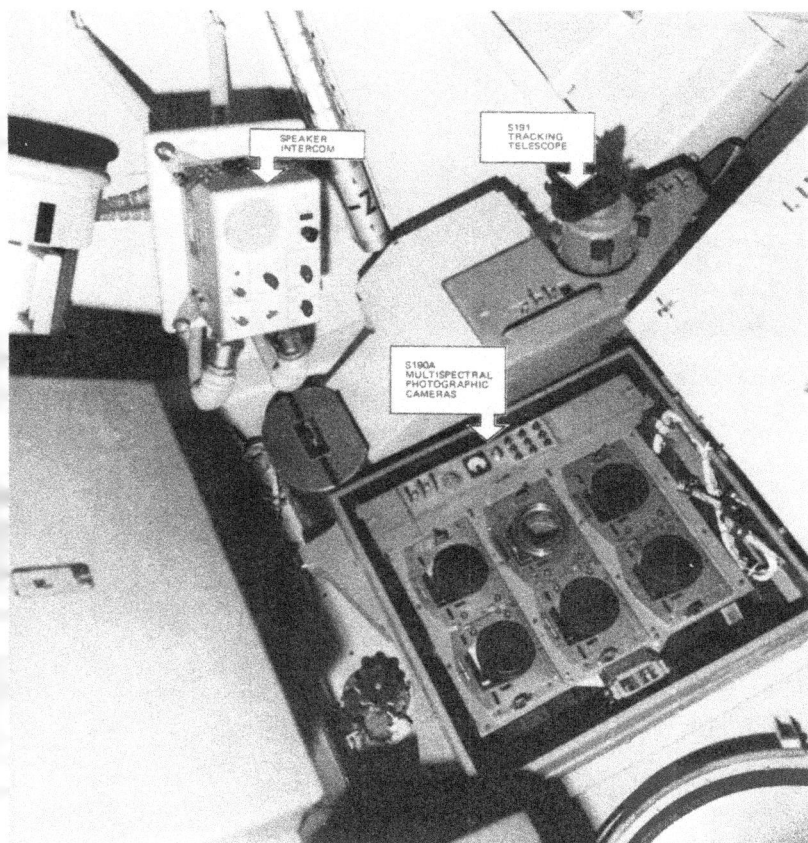

FIGURE 165.—Experiment S190A, Multispectral camera for earth observations.

focal lengths are mounted together and bore-sighted. Wavelength-film combinations are:

Wavelength		Film
Nanometer	*Micrometer*	
500 to 600	0.5 to 0.6	Pan X black and white.
600 to 700	0.6 to 0.7	Pan X black and white.
700 to 800	0.7 to 0.8	Infrared black and white.
800 to 900	0.8 to 0.9	Infrared black and white.
500 to 800	0.5 to 0.8	Infrared color.
400 to 700	0.4 to 0.7	High resolution color.

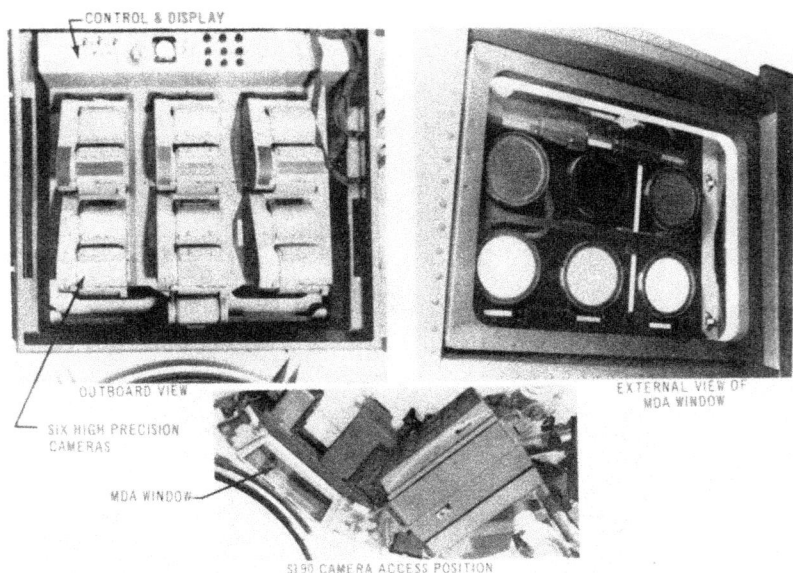

FIGURE. 166.—Experiment S190A, Multispectral camera arrangement in Skylab.

S190B Earth Terrain Camera (Fig. 167)

Project Scientist:
K. Demel
NASA-Lyndon B. Johnson Space Center
Houston, Texas

Objective and Instrumentation:

Obtain high-resolution data of small areas within the fields of view of the Earth Resources Experiment Package sensors to aid in interpretation of data gathered by these sensors (Fig. 168). The camera will offer the first opportunity to obtain high-resolution Earth photography from a manned spacecraft. Ground resolution is expected to be 11 meters (35 ft.) ; ground area coverage will be 109 km (68 statute miles) on a side. The anticipated resolution will be a marked improvement over prior photography obtained on manned flights to date or the photography of the S190A camera.

Crew Involvement:

One crewman will be required to unstow the camera, install it in the Scientific Airlock, operate the controls, and restow the camera. Film has to be loaded in the camera magazine and unloaded after exposure for return to Earth in the Command Module.

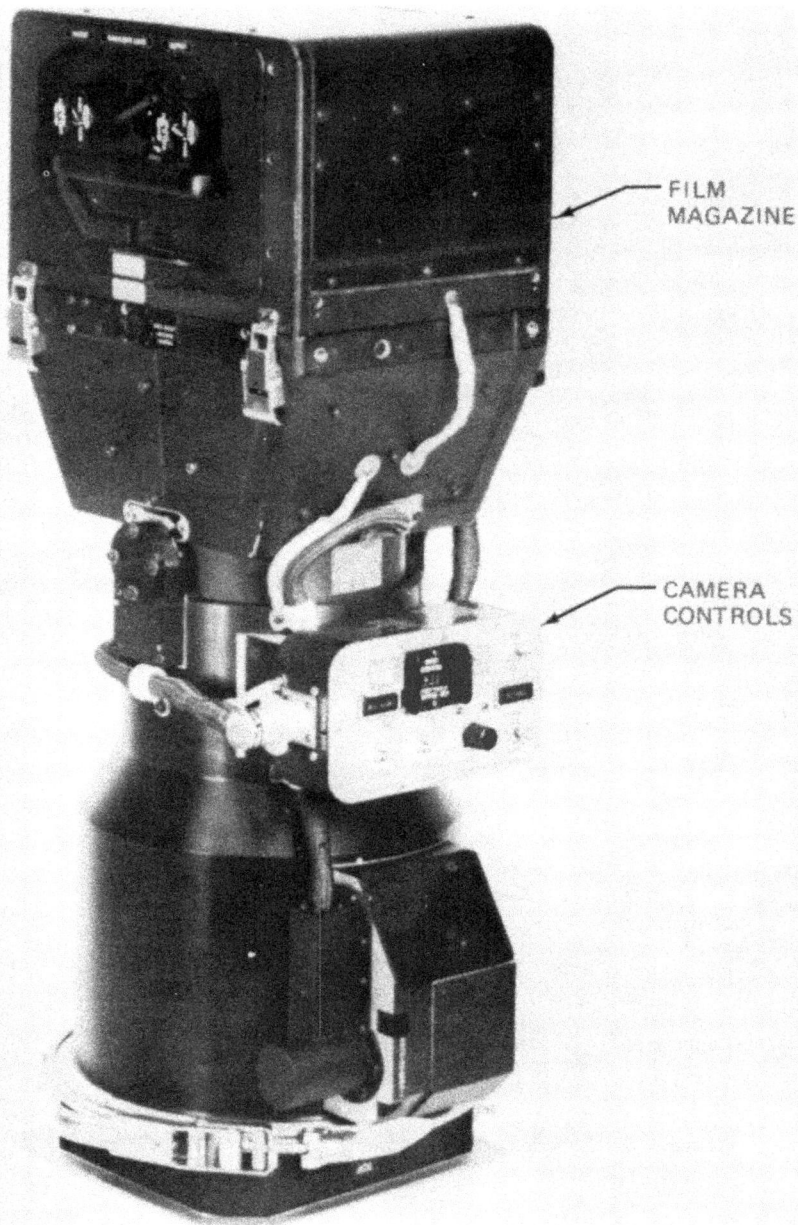

FILM
MAGAZINE

CAMERA
CONTROLS

FIGURE 167.—Experiment S190B, Camera with small field of view for high resolution
pictures of the Earth's surface.

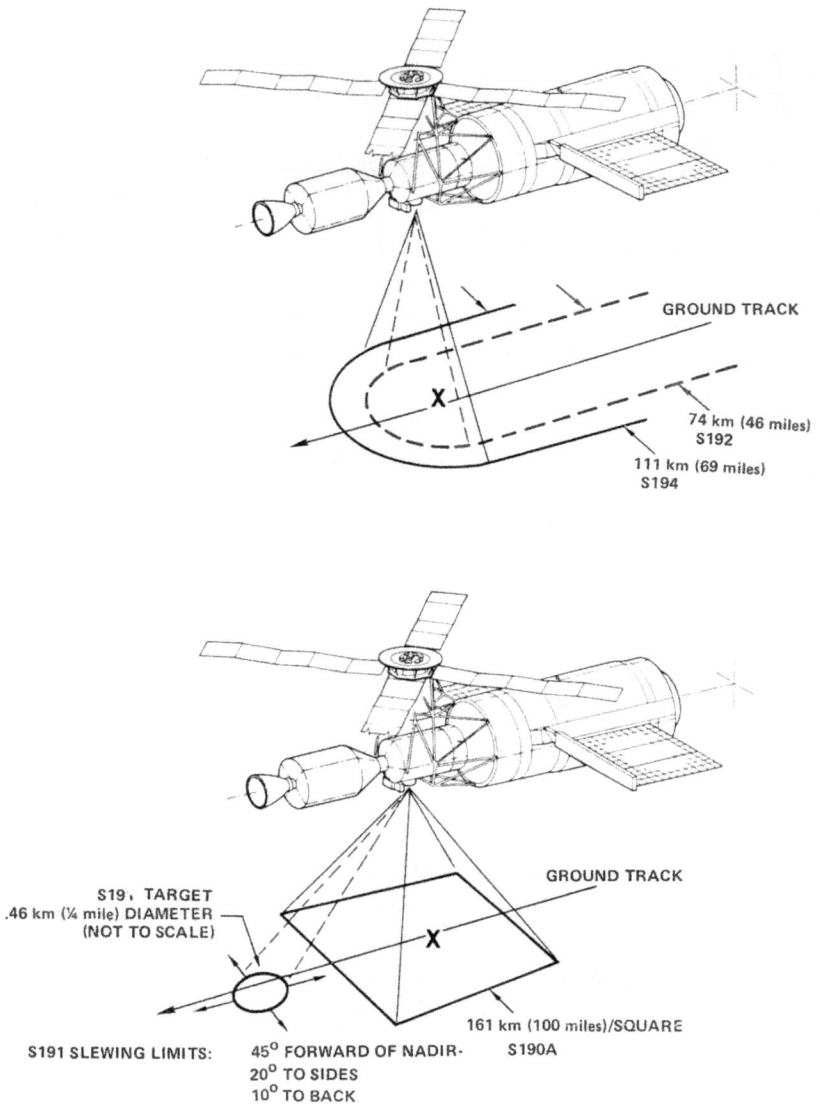

GROUND TRACK

74 km (46 miles)
S192

111 km (69 miles)
S194

S191 TARGET
.46 km (¼ mile) DIAMETER
(NOT TO SCALE)

GROUND TRACK

X

161 km (100 miles)/SQUARE
S190A

S191 SLEWING LIMITS: 45° FORWARD OF NADIR-
20° TO SIDES
10° TO BACK

FIGURE 168.—Experiment S191, Ground coverage of EREP cameras.

b. Infrared Observations

S191 Infrared Spectrometer (Fig. 169)

Project Scientist:
  Dr. T. L. Barnett
  NASA-Lyndon B. Johnson Space Center
  Houston, Texas

Objective and Instrumentation:

Provide data for evaluation of Earth resources sensors for specific regions of the visible and infrared spectra, and for quantitative evaluation of the effects of atmospheric attenuation. These data will enable scientists in various disciplines to evaluate the utility of remote spectrometer sensing from space. Using a filter wheel spectrometer, measurements will be made in the 0.4 to 2.4 and 6.2 to 15.5 micrometer regions (400 to 2400 and 6200 to 15,500 nm). The spectrometer has pointing and tracking capabili-

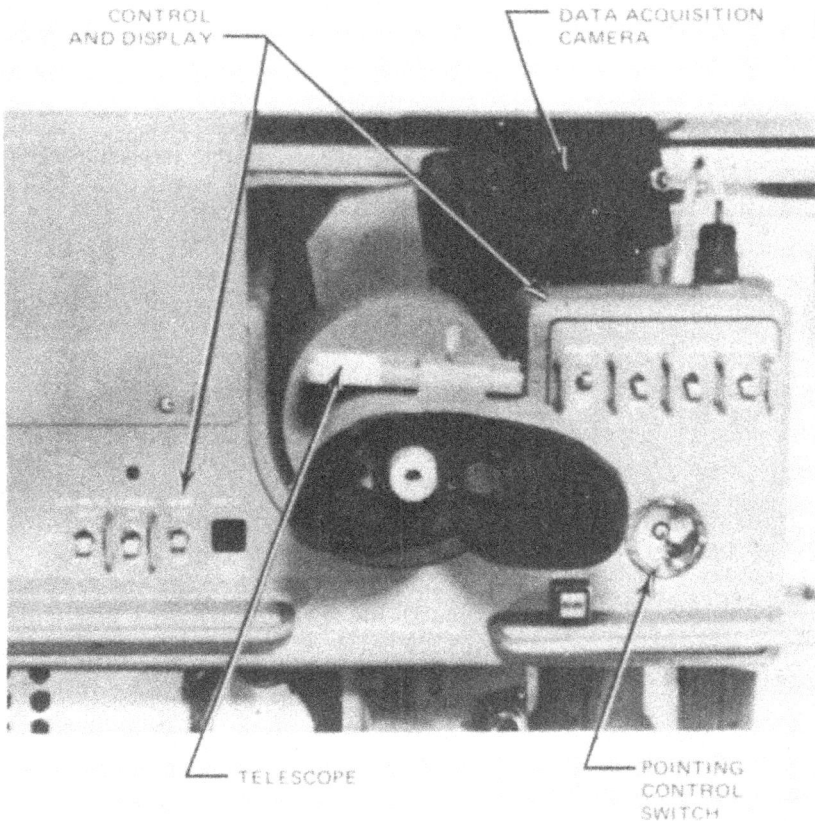

FIGURE 169.—Experiment S191, Infrared spectrometer.

153

ties 45° forward, 10° aft, and 20° to either side, and a field of view of 0.46 km (one-quarter statute mile) in diameter. The astronaut will use a viewfinder and tracking telescope with zoom capability to find interesting ground sites at nadir that usually will be in his field of view for less than a minute.

S192 Multispectral Scanner (Fig. 170)

Project Scientist:
Dr. C. K. Korb
NASA-Lyndon B. Johnson Space Center
Houston, Texas

Objective and Instrumentation:

Provide, and record on tape, sensor signals representing line-scan images of selected test sites. Sensors will respond to 13 bands of reflected and emitted radiations in the visible, near-infrared, and thermal infrared regions of the spectrum (0.4 to 12.5 micrometers). Investigators can then evaluate the usefulness of spacecraft multispectral data for crop identification, vegetation mapping, soil moisture measurements, identification of contaminated areas in large bodies of water, and surface temperature mapping. A dichroic beam splitter is used to separate the incoming radiation into visible to near-IR and thermal IR components. A triple prism spectrometer will disperse the incident radiation in the visible to near-IR region. The instrument has a field of view of approximately 11 degrees and will survey ground swaths 74 km (46 statute miles) wide with a resolution of approximately 79 meters (260 feet).

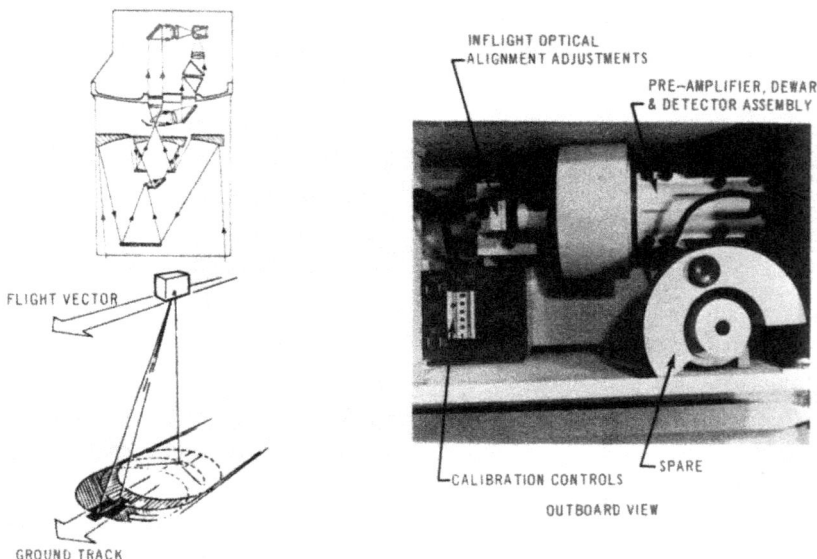

FIGURE 170.—Experiment S192, Multispectral scanning camera.

c. Microwave Studies

S193 Microwave Radiometer/Scatterometer and Altimeter (Fig. 171)

Project Scientist:
D. E. Evans
NASA-Lyndon B. Johnson Space Center
Houston, Texas

Objective and Instrumentation:

Provide simultaneous measurements of reflected radar signals (differential backscattering cross-section) and emitted microwave radiation (brightness

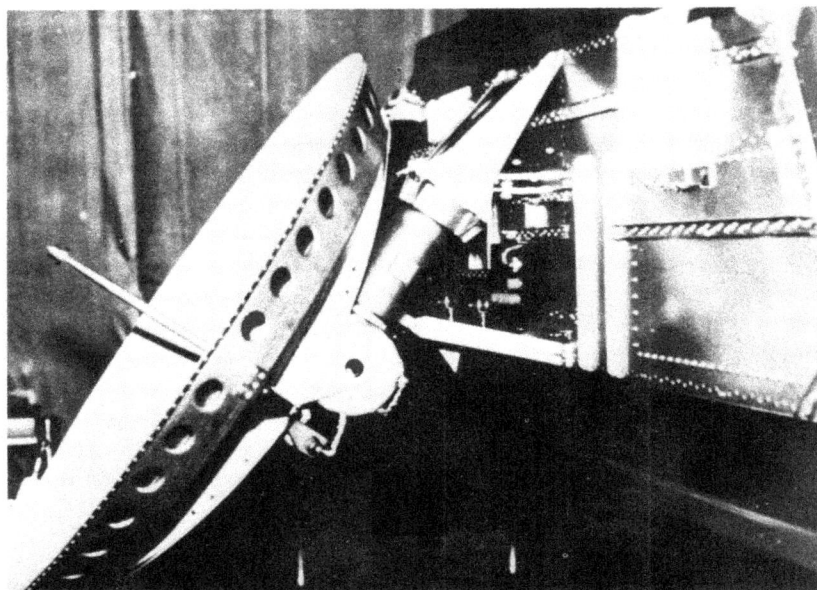

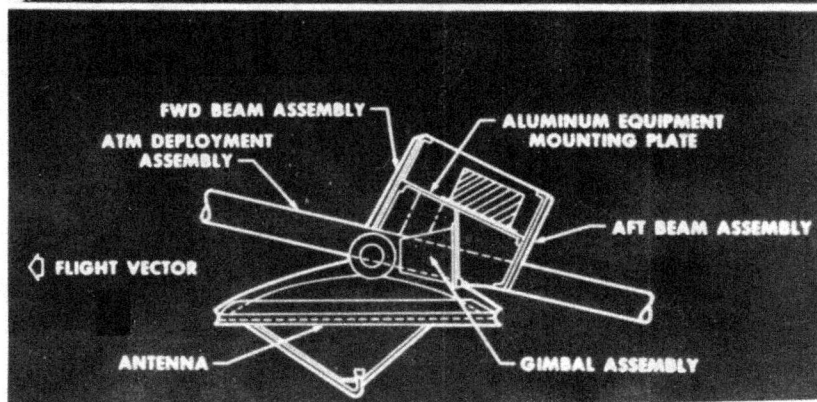

FIGURE. 171.—Experiment S193, Microwave radiometer, scatterometer, and altimeter.

temperature) of land and ocean areas, obtain altimetry data relating sensor response to actual oceanic state. Data will provide information relative to seasonal changes in snow cover and the border between frozen and unfrozen ground, gross vegetation regions and their seasonal changes, flooding, feasibility of measuring soil types and texture, heat output of metropolitan areas, regions of lake and sea ice, and ocean surface characteristics. The transmitter for the radar signals and the receiver for reflected radar signals and emitted microwave radiation will operate at a frequency of 13.9 GHz. Brightness temperature and backscattering will be measured as functions of incidence angle. The instrument will operate in several modes; the maximum forward pointing is 48°, and the maximum side pointing is 48° to either side. The ground area coverage at nadir is 11.1 km (6.9 statute miles) in diameter. The pulse radar altimeter, which shares the antenna assembly with the microwave experiment, will record normal return radar pulses. Their evaluation will provide information about ocean state effects on pulse characteristics. The altimeter has a nadir alignment capability which will give the altimeter a more accurate alignment with the vertical than the vehicle could provide.

S194 L-Band Radiometer (Fig. 172)

Project Scientist:
 D. E. Evans
 NASA-Lyndon B. Johnson Space Center
 Houston, Texas

Objective and Instrumentation:
 Measure and map brightness temperatures of terrestrial surfaces to a high degree of accuracy with a passive radiometer operating at a frequency of

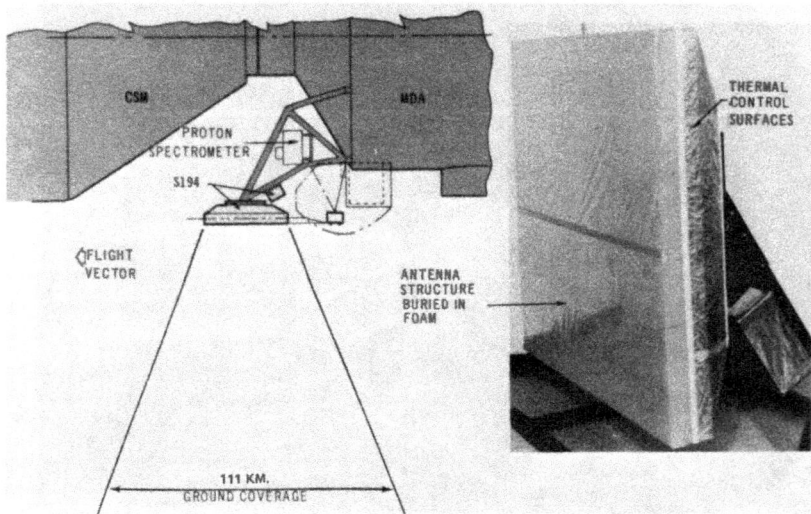

FIGURE 172.—Experiment S194, L-Band radiometer to measure surface temperatures.

1.43 GHz. Measurements of the brightness temperature of the Earth's surface will supplement the S193 experiment. Effects of cloud cover on radiometric measurements can be determined by comparing measurements at both S193 and S194 frequencies (13.9 GHz and 1.43 GHz). The instrument will survey ground swaths 111 km (69 statute miles) wide.

3. LIFE SCIENCES PROJECTS

From the beginning of manned space flight, there has been concern about the ability of man to survive a flight through space and to perform satisfactorily in the space environment. Specific concern has centered around the exposure of the human body to launch accelerations, its adaptation to weightlessness, its ability to withstand reentry loads, and its readaptation to full gravity following the return to Earth.

The first decade of manned space flight was devoted to the preparation of man for the Apollo Program and to his qualification for lunar landing missions. During the Gemini III mission, limited medical experiments were conducted to study man's physiological reactions during a two-week mission. The other Gemini and Apollo flights were used for studies of physiological effects on man through pre- and postflight medical experiments.

Before long-duration programs of explorations and operations in space can be undertaken, man's viability and usefulness under space conditions must be further assured. This can only be accomplished through careful quantitative studies of man's physiological, psychological, and social adjustments as they occur during flight. Limiting influences exercised by the space environment on the capabilities of crew members must be studied, and proper levels of performance for any given time during a flight must be established. These studies will result in time profiles of the adaptation of men to space conditions, and they will show whether long-term adjustments eventually lead to new stable levels, or whether the need for continual adjustments threatens to exceed man's reserve capacity for meeting stress. Even if the crew members do successfully adapt to space conditions, the return to Earth involves an additional adaptive change about which more must be learned.

The Skylab Program offers the first opportunity to study these questions in depth. The 28- and 56-day missions are long enough for a study of acute effects which could threaten man's safety, and also for the observation of slower biological processes. The biomedical experiments on Skylab have been designed to study the suspected changes and to investigate the basic mechanisms involved in these changes. The experimental investigations are much more comprehensive than previous investigations which served only medical safety monitoring purposes. Medical safety monitoring will be performed operationally on Skylab by known and fully tried bioinstrumentation, medical techniques and procedures.

The Skylab medical program represents an intensive study of normal, healthy men and their reactions to the stresses of space flight. Seldom has such a comprehensive examination been performed in ground-based studies and never under the unusual stresses of prolonged space flight. As an additional benefit, preparing and conducting these multi-man extended missions will lead to advances in Earth-based medicine in such areas as non-invasive biosensors (medical probes which do not have to be inserted into the body) and biotelemetry where significant contributions to medical diagnosis and treatment are expected.

A basic set of biomedical data has been collected as a safety monitoring procedure on all manned flights of the Mercury, Gemini and Apollo programs. Heart and respiration rates, and at times body temperature and blood pressure were recorded. These data were supplemented by a variety of pre- and postflight measurements of such factors as exercise capability, cardiovascular stress response, hematological-biochemical changes, immunology studies, and microbiological evaluations. In the Gemini program, medical experiments of limited scope were conducted in flight to investigate the time course of the changes which had been noticed before and after previous missions.

The following physiological effects of space flight on man have been observed:

A consistent loss of body weight; a small and inconsistent loss of bone calcium and muscle mass; and generally after return to Earth, a reduction in the ability of blood vessels to actively distribute blood to those parts of the body that need it (orthostatic intolerance).

These effects completely reversed themselves within a few days after return to Earth. So far, they have shown no consistent relation to flight duration (up to 14 days). However, some concern remains that continued exposure to flight conditions on extended missions could significantly reduce man's effectiveness in space and increase the difficulty of re-adapting to the gravity conditions on Earth or on another celestial body.

Each manned mission in the U.S. space program was built upon the cumulative experience of preceding flights. Skylab will expose more men, in a larger spacecraft, with more varied activities, and for longer times to the weightlessness of orbital flight than any previous space project. It will allow more thorough evaluation of biomedical observations under extended periods of zero gravity, and it will use more rigorous evaluation techniques, than has previously been possible (Fig. 173).

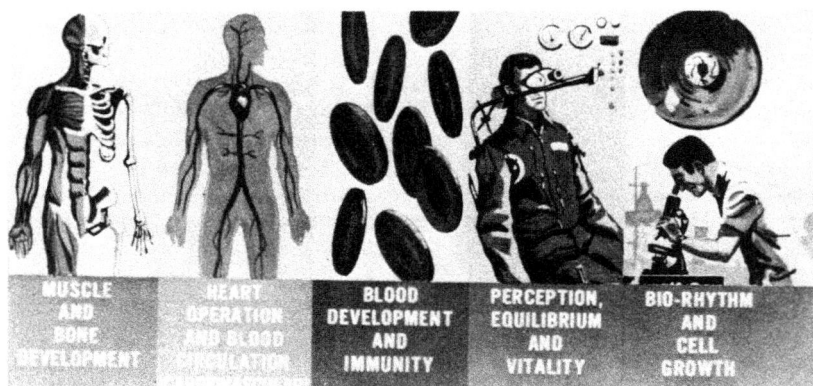

FIGURE 173.—Medical studies carried out on Skylab and on the ground.

The Skylab biomedical program will cover four areas:

• The project will achieve extended stay times of nine men in space, three at a time, with the associated operational medical monitoring and the observations of crew performance in a wide variety of scientific and operational tasks.

• The medical experiments are designed to investigate in depth those physiological effects and their time courses which were observed in previous flights.

• The biology experiments are designed to study fundamental biological processes which may be affected by the weightless environment.

• The biotechnology experiments are directed toward advancing the effectiveness of man-machine systems in space operations and improving the technology of space-borne bioinstrumentation.

The knowledge and experience gained from all four parts of the program will help establish criteria for incremental increases in the duration of future manned missions after the 28- and 56-day Skylab flights.

Two animal experiments on Skylab (S071, S072) will help determine whether the force of gravity may have an influence on the regulators of some of the fundamental rhythms in living organisms. A small colony of mice and a sampling of pupae of the vinegar gnat will be flown on Skylab and observed for changes in specific life cycles under the zero-gravity condition of space.

M071, Mineral Balance

Principal Investigator:
  G. Donald Whedon, M.D.
  National Institutes of Health,
  Washington, D.C.
Co-Investigator:
  Leo Lutwak, M.D.
  Cornell University
  Ithaca, N.Y.
Principal Coordinating Scientist:
  Dr. Paul C. Rambaut,
  NASA-Lyndon B. Johnson Space Center
  Houston, Texas.
Development Center:
  JSC; Integration Center: MSFC

Objective:

Collect data for a predictive understanding of the effects of space flight on the muscle and skeletal system by measuring the day-to-day gains or losses of pertinent biochemical constituents.

The following data are to be collected: daily body weight; accurate food intake (quantity and composition); accurate fluid intake; volume of a 24-hour urine output; samples of pooled 24-hour urine output; and preflight, inflight and postflight blood samples taken for analysis. Also, all feces and all vomitus (if any) will be collected, weighed, processed, and stored for return and postflight analysis.

Urine will be analyzed for calcium, phosphorus, magnesium, sodium, potassium, chlorine, nitrogen, urea, hydroxyproline, and creatinine. Feces will be analyzed for calcium, sodium, phosphorus, magnesium, potassium, and nitrogen. Blood will be analyzed for calcium, phosphorus, magnesium, alkaline phosphotase, sodium, potassium, total protein, glucose and hydroxproline, creatinine, chloride, and electrophoretic pattern.

Instrumentation:

All instruments used in this experiment are parts of other systems. They will be described in the appropriate sections. These instruments include the following:

Urine Measurement and Collection System (a part of the Habitability Support System).

Fecal Collection System (a part of the Habitability Support System).

Specimen Mass Measurement Device (a part of M074).

Body Mass Measurement Device (a part of M172).

Food system.

Inflight blood collection equipment.

M073, Bioassay of Body Fluids (Fig. 174)

Principal Investigator:
Dr. Carolyn S. Leach
NASA-Lyndon B. Johnson Space Center
Houston, Texas
Development Center:
JSC; Integration Center: MSFC

Objective:

Assess the effect of space flight on endocrine-metabolic functions including fluid and electrolyte control mechanisms. Collect the following data: daily body weight; accurate food intake (quantity and composition);

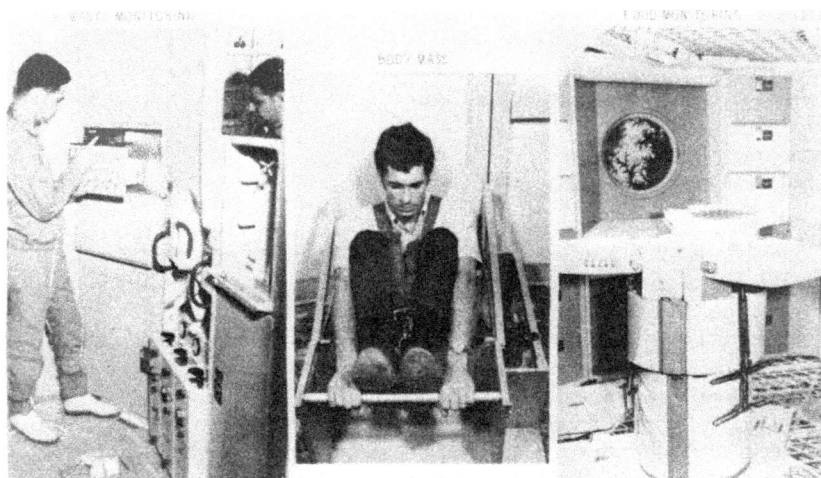

FIGURE 174.—Experiments to monitor mass changes of astronauts.

accurate fluid intake; volume of a 24-hour urine output; samples of pooled 24-hour urine output (collected and processed inflight for return and postflight analysis); and preflight, inflight, and postflight blood samples taken for analysis.

Urine will be analyzed for sodium, potassium, aldosterone, epinephrine, norepinephrine, antidiuretic hormones (ADH), urine osmolality, hydrocortisone, total body water, and total and fractional ketosteroids. Blood will be analyzed for renin, sodium, potassium, chloride, plasma osmolality, extracellular fluid volume (ECF), parathyroid hormone, thyrocalcitonin, thyroxine, adrenocorticotropic hormone (ACTH), hydrocortisone, and total body water.

Instrumentation:

All instruments used in this experiment are parts of other systems; they include the following:

Urine Measurement and Collection System (a part of the Habitability Support System).

Specimen Mass Measurement (a part of M074).

Body Mass Measurement (a part of M172).

M074, Specimen Mass Measurement (Fig. 175)

Principal Investigator:

William E. Thornton, M.D.

Astronaut

NASA-Lyndon B. Johnson Space Center

Houston, Texas

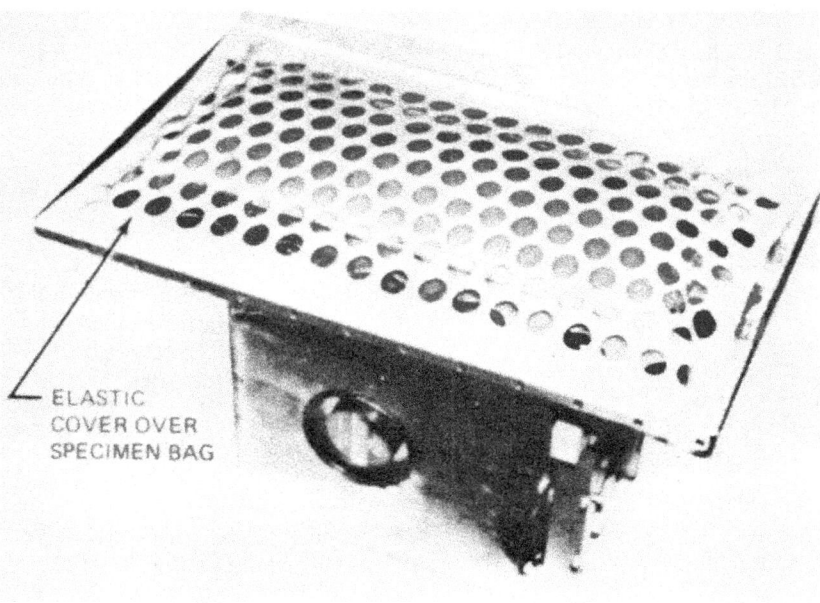

ELASTIC
COVER OVER
SPECIMEN BAG

FIGURE 175.—Experiment M074, Weighing Device to determine masses of specimens.

Co-Investigator:
John W. Ord, Colonel, Medical Corps
USAF Hospital, Clark AF Base, Philippine Islands
Development Center:
JSC; Integration Center: MSFC

Objective:

Demonstrate the capability of accurately determining masses from 50 to 1000 grams in a zero-gravity environment. Determine the masses of feces, vomitus, and food residue generated in flight.

Data to be collected include the following: preflight calibration of the Specimen Mass Measurement Device (SMMD), measurement of known masses three times during each Skylab mission, and Skylab environmental temperature. The SMMD will also be used to determine the masses of feces, vomitus, and food residue.

Instrumentation:

The SMMD utilizes the inertial property of mass instead of the gravitational force to determine mass. Basically, the SMMD consists of a spring-mounted tray. The oscillatory period of the spring is a function of the amount of mass on the tray. The spring's period is measured electro-optically, and this measurement is electronically converted to a direct mass read-out.

M078, Bone Mineral Measurement

Principal Investigator:
John M. Vogel, M.D.
U.S. Public Health Service Hospital
San Francisco, California
Co-Investigator:
Dr. John R. Cameron
University of Wisconsin Medical Center
Madison, Wisconsin
Development Center:
JSC

Objective:

Assess the effects of the spaceflight environment on the occurrence and degree of bone mineral changes in the left heel and forearm (radius) by measuring bone masses before and after Skylab flights. These measurements will indicate the degree of calcium deposition (calcification) in the bones. Normal chemical activity in bones is stimulated by the pulling force of attached muscles and by gravitational forces acting upon the body. Both forces are altered during weightless flight.

Instrumentation:

Measurements of bone masses will be taken only on the ground, before and after flight. A scanning probe, using the soft gamma radiation of the radioactive isotope iodine-125, a gamma ray detector, and a multichannel analyzer will take radiograms of the bones; by comparing postflight with preflight pictures, the changes can be determined. Calibration will be achieved by comparing the bone absorption, mostly due to calcium, with the absorption in test layers of known thickness and composition. Measurements will be taken of all crew members and of control group members.

M092, Lower Body Negative Pressure (Figs. 176, 177, 178, 179)

Principal Investigator:
  Robert L. Johnson, M.D.
  NASA-Lyndon B. Johnson Space Center
  Houston, Texas
Co-Investigator:
  John W. Ord, Colonel, Medical Corps
  USAF Hospital, Clark AF Base, Philippine Islands
Development Center:
  JSC; Integration Center: MSFC

Objective:
  Determine the degree and the time course of cardiovascular adaptation under zero-gravity conditions. Provide data on cardiovascular changes

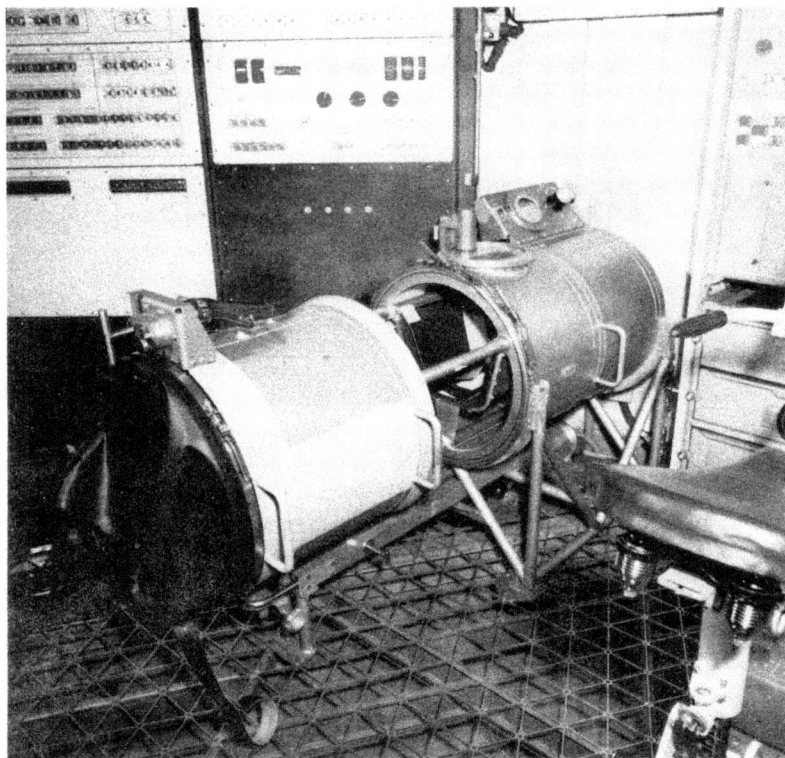

FIGURE 176.—Experiment M092, Instrument to apply negative pressure to the lower part of the human body (instrument in open position).

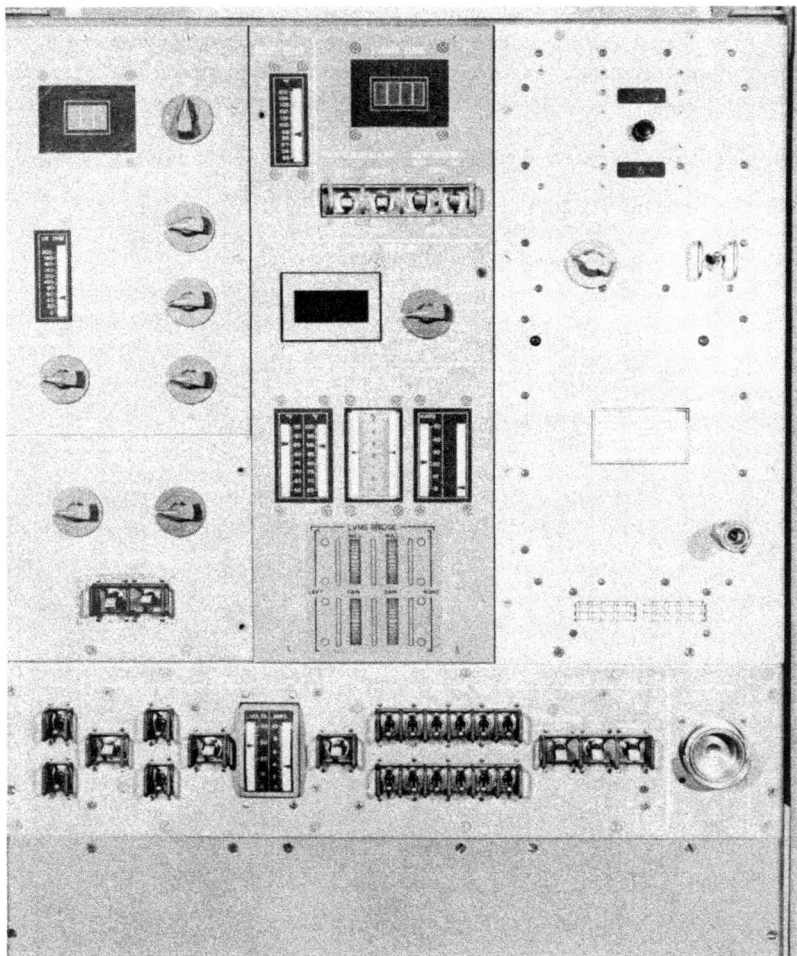

FIGURE 177.—Control Panel to support medical experiments.

for correlation with preflight and postflight measurements. Collect inflight data for predicting the impairment of physical capacity and the degree of orthostatic intolerance to be expected upon return to Earth (orthostatic intolerance is the inability of the organism to distribute the blood properly, and under the proper pressures among the different organs and places in the body when the body assumes an erect position in a gravitational field). Data to be collected are blood pressure, heart rate, body temperature, vectorcardiogram, lower body negative pressure, leg volume changes, and body mass.

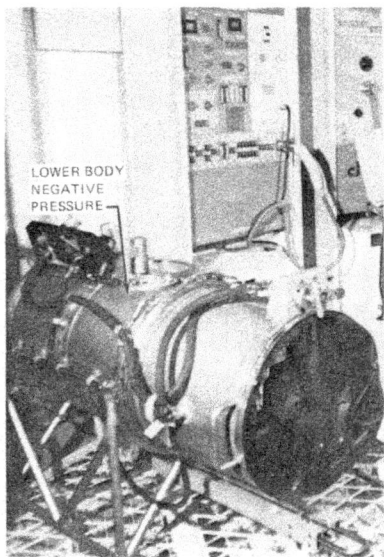

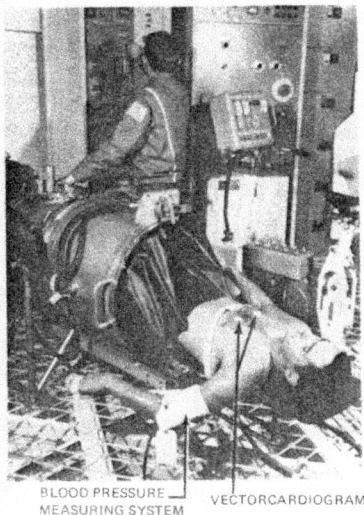

FIGURE 178.—Experiment M092, Lower Body Negative Pressure instrument training on the ground.

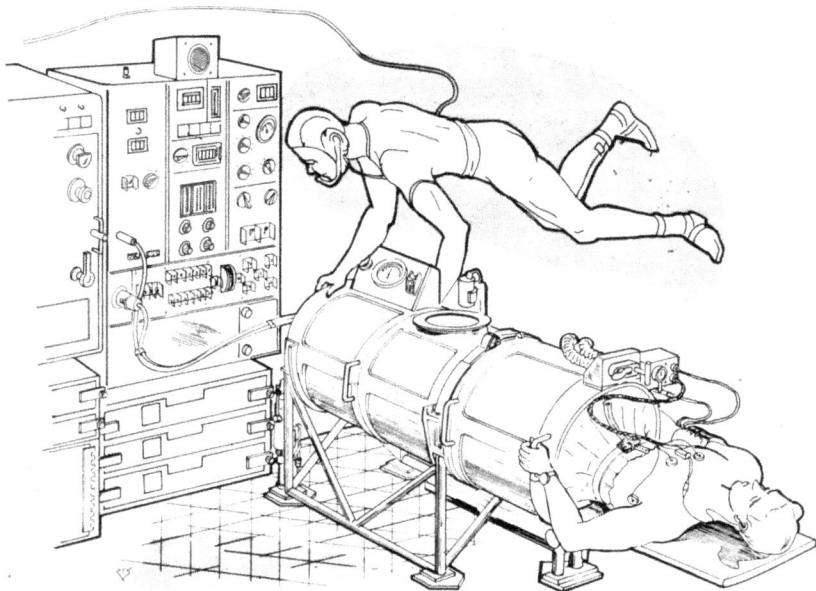

FIGURE 179.—Experiment M092, Lower Body Negative Pressure instrument in operation on Skylab.

The Lower Body Negative Pressure (LBN) experiment imposes a slight reduction of external pressure to the lower half of the body. This "negative" pressure (negative with respect to the environment of the upper half of the body) will have a blood pooling effect in the lower part of the body, similar to the effects of the normal hydrostatic pressure of the blood column in a person standing upright in the Earth's gravity field. The experiment will test how the cardiovascular system reacts to a controlled amount of blood pooling during weightless flight.

Instrumentation:

The LBNP experiment utilizes three basic units: (1) a cylindrical tank with a waist seal which encloses the lower half of the astronaut under test. The pressure in the tank can be lowered by as much as 15 to 20 percent below the ambient cabin pressure, thus exposing the lower body to a controlled series of negative pressures. (2) a leg volume measuring system (Leg Volume Plethysmograph) which senses the expansion of the legs by measuring the circumference of each leg at the level of the calf muscle. The amount of expansion is a measure of the amount of blood pooling in the legs. (3) the Blood Pressure Assembly, consisting of a pressure cuff attached to the upper arm, a microphone to pick up the sounds of blood flow, and the necessary electronic systems. A programming unit cycles the pressure cuff automatically. Recording and calibrating is accomplished through the Experiment Support System.

The experiment uses the Vectorcardiogram equipment from M093 and the Body Temperature Measuring System from M171. It will be performed on each astronaut every three days. The entire experiment takes about 60 minutes to perform; an attending astronaut is needed to assist the subject.

M093, Vectorcardiogram (Figs. 180, 181)

Principal Investigator:
Newton W. Allebach, M. D.
USN Aerospace Medical Institute
Pensacola, Florida
Co-Investigator:
Raphael F. Smith, M. D.
School of Medicine
Vanderbilt University
Nashville, Tennessee
Development Center:
JSC; Integration Center: MSFC

Objective:

Measure the activity of the heart by recording electric signals (vectorcardiographic potentials) of each astronaut during preflight, inflight, and postflight periods to obtain information on changes in heart functions induced by the flight conditions. Vectorcardiograms will be taken at regu-

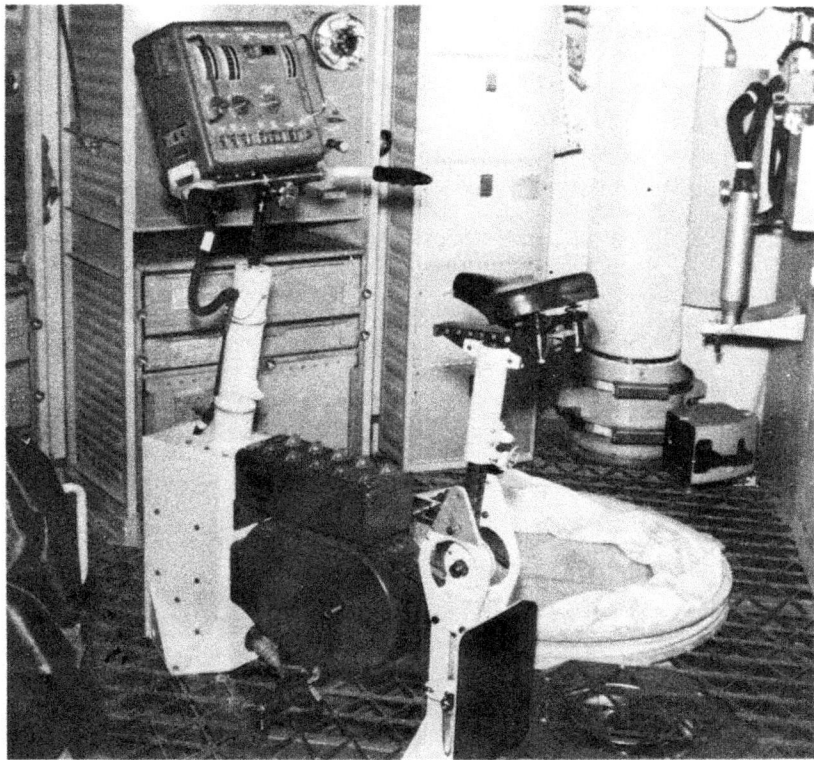

FIGURE 180.—Experiment M093, Ergometer (wheelless bicycle) to record vector-cardiograms during controlled physical exercise.

lar intervals throughout the mission while the crewmen are at rest, and before, during, and after specific exercise periods with a bicycle ergometer (part of Experiment M171). This instrument will enable an astronaut to exercise at selected levels of energy consumption.

Instrumentation:

The Vectorcardiogram system consists of a harness with eight electrodes, a signal-shaping network (Frank Lead Network), calibration and timing circuits, and three Electrocardiogram (ECG) signal conditioner channels. The system provides three ECG signals and a heart rate signal to the spacecraft system and to the Experiment Support System (ESS). The ESS will provide power, additional signal conditioning, and recording facilities.

500-721 O - 73 - 12

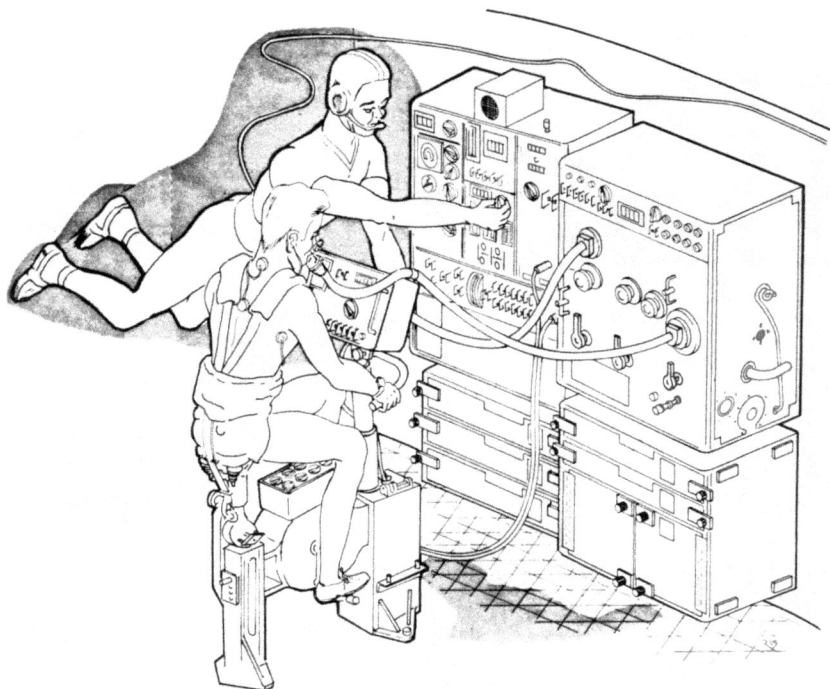

FIGURE 181.—Experiment M093, Recording of vectorcardiograms during physical exercise in orbit.

M111, Cytogenic Studies of the Blood

Principal Investigator:
  Lilliam H. Lockhart, M.D.
  University of Texas
  Medical Branch
  Galveston, Texas
Co-Investigator:
  P. Carolyn Gooch
  Brown & Root-Northrop
  Houston, Texas
Development Center:
  JSC

Objective:

Determine the frequency of chromosome changes in the peripheral blood leukocytes of the Skylab crew members before and after flight. Correlate the results with the radiation doses received by the astronauts ("in vivo" radiation dosimetry). Acquire data that will add to the findings of other Skylab cytologic and metabolic experiments to determine the genetic consequences of long duration space travel on man.

Chromosomes, located in the nuclei of cells, provide the basis for control of most of the biochemical functions and activities within an organism. Their internal structures are susceptible to change under the influence of radiation, chemical reagents, and some other environmental factors, possibly including weightlessness.

Instrumentation:

Periodic blood samples will be taken before and after the flight, beginning one month before and ending three weeks after recovery. The leukocytes will be placed in a short-term tissue culture. During the first cycle of cell division in the isolated cultures, standard chromosome preparations of the leukocytes will be made.

The leukocytes from the cell culture will be removed during metaphase and "fixed." A visual analysis will be performed which involves counting the chromosomes, the number of breaks, and possibly the types of breaks, and then comparing the identifiable chromosome forms with groups of chromosomes comprising the normal human complement.

This experiment will provide chromosome aberration frequencies as they are found postflight in nine men. The relation between radiation dose experienced by each man and the number of chromosome breaks will be studied on the basis of these data.

M112, Man's Immunity, In-Vitro Aspects

Principal Investigator:
Stephen E. Ritzman, M.D.
University of Texas Medical Branch
Galveston, Texas
Co-Investigator:
William C. Levin, M.D.
University of Texas Medical Branch
Galveston, Texas
Development Center:
JSC; Integration Center: MSFC

Objective:

Determine any changes in man's cell chemistry that may result from prolonged exposure to weightlessness. Study the changes in humoral and cellular immunity as reflected by the concentration of plasma and blood cell proteins, blastoid transformations, and synthesis of ribonucleic (RNA) and desoxyribonucleic (DNA) acids by the lymphocytes.

An organism's ability to combat infections or to repair injured tissues may be influenced by the lack of gravity as a consequence of a change in cell chemistry caused by zero gravity.

Instrumentation:

Data reflecting normal cell metabolism will be obtained 21, seven, and one day before launch from the crewmen, and from a control group composed of three men physically similar to the crewmen. This group will also serve as the ground control group during the flight. Inflight blood samples will be taken four times from each crewman during the first mission and eight times from each crewman during the second and third missions. Seven days and 21 days after recovery, samples again will be taken from each crewman.

Blood will be analyzed for kinetics of lymphocyte RNA and DNA, RNA and DNA distribution in lymphocytes, observation of blastoid formation, lymphocyte morphology and antigen response, lymphocyte functional response to antigen quantitation of plasma constituents, presence of immunoglobulins, albumin and globulin concentration, and total plasma protein. The Inflight Blood Collection System will provide the capability to draw venous blood and to centrifuge the samples for preservation. The blood samples will be frozen and returned to Earth for postflight analysis. The data from preflight, inflight, and postflight samples will be compared to detect any significant changes caused by the conditions of orbital flight.

M113, Blood Volume and Red Cell Life Span (Fig. 182)

Principal Investigator:
 Phillip C. Johnson, M.D.
 Baylor University
 College of Medicine
 Houston, Texas
Development Center:
 JSC; Integration Center: MSFC

Objective:

Determine the effects of weightlessness on the blood plasma volume and the red blood cell population. Particular attention is to be paid to changes in total mass of red cells, red cell destruction, red cell life span, and red cell production rate.

Red blood cells transport oxygen from the lungs to all parts of the body. Decreases in the total mass of red cells will necessitate increased heart and breathing rates.

Instrumentation:

This experiment has four parts; in each, a different radioisotope tracer will be injected into crewmen's veins and into veins of a control group with similar physical characteristics on the ground.

The site of red blood cell (RBC) production in the mature adult is the marrow of membranous bones (e.g. sternum and vertebrae). The rate of production is dependent on metabolic demands and on the current red cell population. The rate of RBC production will be measured quantitatively by injection of a known quantity of a radioactive ion tracer into crew members.

Since the rate of RBC production acts with RBC loss to increase or decrease the total RBC mass present at a given time, any changes in the rates of RBC production and destruction will be necessarily reflected in the red cell mass. Such changes in red cell mass will be measured and analyzed in the flight crew members by injection of red cells tagged with radioactive chromium (in the form of sodium chromate).

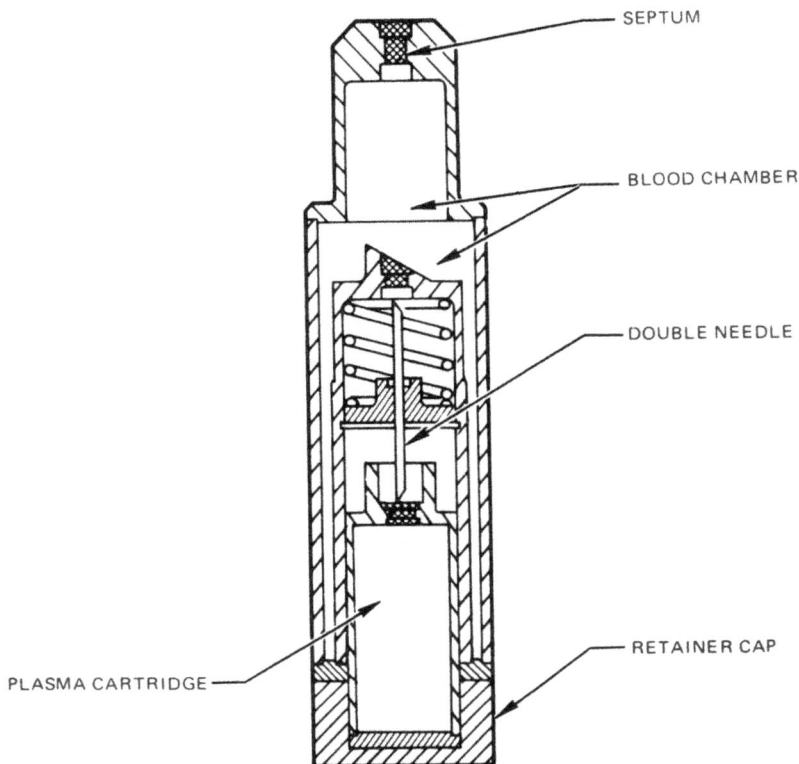

FIGURE 182.—Experiment M113, Automatic syringe to take blood samples of astronauts during orbital flight.

To determine selective age-dependent RBC destruction and mean red cell life span, glycine tagged with radioactive carbon will be injected into a superficial arm vein of each crew member and control subject.

Finally, plasma volume changes will be measured by adding a known amount of radioiodinated human serum albumin to each crew member's blood.

The Inflight Blood Collection System will provide the capability to draw venous blood and to centrifuge the samples for preservation; the blood samples will be frozen and returned to Earth for postflight analysis.

Blood samples of each crewman will be taken preflight (21, 20, 14, seven and one days before launch); inflight (four times during the first and eight times during the second and third manned mission); and postflight (recovery day, one, three, seven, 14 and 21 days after recovery).

M114, Red Blood Cell Metabolism

Principal Investigator:
  Charles E. Mengel, M.D.
  University of Missouri
  School of Medicine
  Columbia, Missouri
Development Center:
  JSC; Integration Center: MSFC

Objective:

Study the effects of gravity on the membrane and the metabolism of the human red blood cell. Determine whether any metabolic changes or membrane modifications occur as a result of exposure to the space flight environment. This experiment will complement Experiment M113.

Instrumentation:

Blood samples of each crewman will be taken preflight (21, seven, and one days before launch), inflight (four times during the first and eight times during the second and third manned missions), and postflight (recovery day, one, and 14 days after recovery).

Blood will be analyzed for methemoglobin, glyceraldehyde-6-phosphate, dehydrogenase, phosphoglyceric acid kinase, reduced gluthathione, adenosine triphosphate, gluthathione reductase, lipid peroxide levels, acetylcholinestecase, phosphofructokinase, 2,3-diphosphoglycerate, and hexokinase.

The Inflight Blood Collection System will provide the capability to draw venous blood and to centrifuge the samples for preservation. The blood samples will be frozen and returned to Earth for postflight analysis.

M115, Special Hematological Effects (Fig. 183)

Principal Investigator:
  Dr. Stephen L. Kimsey
  Craig L. Fischer, M.D.
  NASA-Lyndon B. Johnson Space Center
  Houston, Texas
Development Center:
  JSC; Integration Center: MSFC

Objective:

Examine critical physiological blood parameters relative to a stable state of equilibrium between certain blood components and evaluate the effects of weightlessness upon these parameters. Provide other data on blood and blood circulation which will assist in the interpretation of hematology and immunity experiments (M111 series) and of nutrition and musculoskeletal function experiments (M071 series). The red blood cell represents a model system for the evaluation of physiological changes that might occur in man during prolonged exposure to weightlessness.

Blood studies made on Gemini and Apollo astronauts have shown that changes in red cell mass, blood constituents, and the fluid and electrolyte balance can be expected as a result of the space environment.

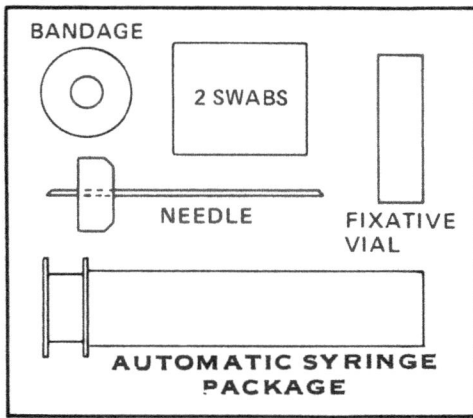

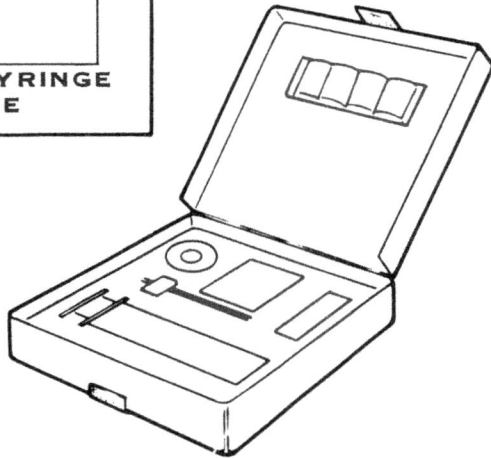

FIGURE 183.—Experiment M115, Blood Collection Kit to sample blood of astronauts during orbital flight.

Instrumentation:

Blood samples of each crewman will be taken preflight (21, 14, seven, and one days before launch), inflight (four times on the first and eight times on the second and third manned missions), and postflight (recovery day, one, three, seven, 14, and 21 days after recovery). Blood will be analyzed for sodium, potassium, single cell hemoglobin, red blood cell hemoglobin, RNA, protein distribution, hemoglobin characterization, electrophoretic mobility, red blood cell age profile, red blood cell electrolyte distribution, membrane and cellular ultrastructure, acid and osmotic fragility, critical volume, volume distribution, red blood cell count, white blood cell count, differential white cell count, microhematocrit, platelet count, hemoglobin, and reticulocyte count.

The Inflight Blood Collection System will provide the capability to draw venous blood and to centrifuge the samples for preservation. The blood samples will be frozen and returned to Earth for postflight analysis.

M131, Human Vestibular Function (Fig. 184)

Principal Investigator:
  Ashton Graybiel, M.D.
  USN Aerospace Medical Institute
  Pensacola, Florida
Co-Investigator:
  Dr. Earl F. Miller
  USN Aerospace Medical Institute
  Pensacola, Florida
Development Center:
  JSC; Integration Center: MSFC

Objective:

Examine the effects of weightlessness on the vestibular system, i.e. the system of semicircular canals of the ear which provides the perception of balance and orientation. Determine any changes in man's sensitivity to motion and rotation and any variations in his ability of coordination under prolonged weightlessness.

Test the astronauts' susceptibility to motion sickness in the Skylab environment, acquire data fundamental to an understanding of the functions of human gravity receptors under prolonged absence of gravity, and test for changes in the sensitivity of the semicircular canals. The following data are to be collected: threshold perception of rotation, motion sickness symptoms caused by out-of-plane head motions while being rotated, and ability of a crewman to determine his orientation with respect to spacecraft reference points without visual cues. Data will be collected before, during, and after flight.

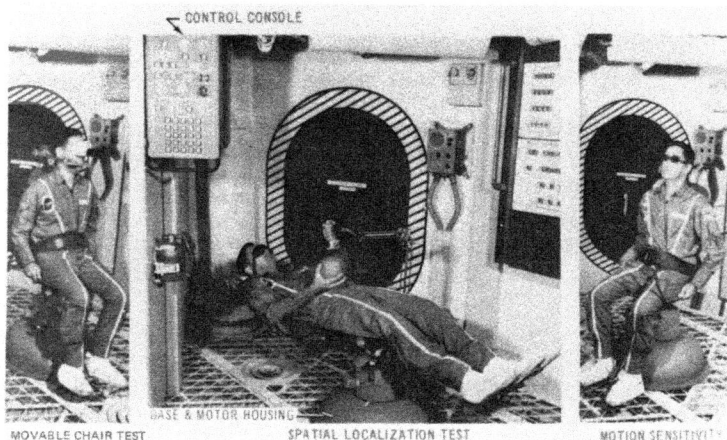

FIGURE 184.—Experiment M131, Study of human vestibular functions on a litter chair, and by marking directions on a magnetic sphere.

Instrumentation:

The inflight equipment includes:

Rotating Litter Chair—This chair is a framed seating device which is convertible for operation in either a rotating or a tilt litter mode.

Drive Motor for Chair Rotation—This motor has the capability of rotating the seated subject within the limits of 1 to 30 rpm at an accuracy of ± 1 percent.

Control Console—The console contains mode selector, speed selector, tachometer, indicators, timers, other devices for control, and a response matrix for coding a subject's response to the rotational tests.

Otolith Test Goggle—This device is used to measure the visual space orientation in two dimensions. It provides the visual target for the oculogyral illusion test.

Custom Bite Boards—The bite boards are used to hold the otolith test goggle precisely and comfortably in position over the observer's eyes.

Reference Sphere and Magnetic Pointer with Readout Device—These devices are used for measuring spatial localization using nonvisual clues. A magnetic pointer is held against the sphere and moved by the subject. This will determine the subject's ability to judge his orientation. The pointer position is measured by the three-dimensional readout device.

M133, Sleep Monitoring (Fig. 185)

Principal Investigator:

James D. Frost, Jr., M.D.
Baylor College of Medicine
Houston, Texas

Development Center:

JSC; Integration Center: MSFC

Objective:

Determine the quantity and quality of an astronaut's sleep during long periods of weightlessness through an analysis of electroencephalographic (EEG) and electro-oculographic (EOG) activities. This information will complement other investigations concerning reactions of the central nervous system under space flight conditions. The following data are to be collected: preflight EEG and EOG data of a crewman for three consecutive nights of sleep, periodical inflight EEG and EOG data throughout a crewman's sleep period, and postflight sleep EEGs and EOGs on approximately the first, the third, and the fifth day after recovery.

Instrumentation:

One of the astronauts, selected for this experiment, will wear a fitted cap during his sleep periods with electrodes for electroencephalographic measurements of brain waves (EEG signals), with accelerometers to record motions of the head, and with electrodes near one eye (electro-oculograph) to sense rapid movements of the eyeball during sleep. Signals from these senors, recorded on magnetic tape and analyzed after return to Earth, will permit conclusions regarding the depth and length of the sleep stages. Signals will also be telemetered to the ground station in near real time. Inflight recordings will be compared with preflight and postflight observations made on the same crew member with the same sensors.

FIGURE 185.—Experiment M133, Monitoring of sleep and sleep reactions of astronaut.

M151, Time and Motion Study (Fig. 186)

Principal Investigator:
  Dr. Joseph F. Kubis
  Fordham University
  Bronx, New York
Co-Investigator:
  Dr. Edward J. McLaughlin
  NASA Headquarters, OMSF
  Washington, D.C.
Development Center:
  JSC; Integration Center: MSFC

Objective:

Observe astronauts in motion. Compare their mobility and dexterity in various activities under weightlessness with their mobility and dexterity in similar activities under Earth conditions. Evaluate their zero-gravity behavior for designs and work programs of future spacecraft.

The following tasks have been selected for this observation:

Study the locomotion of crewmen as they move in the zero-gravity environment with and without loads.

Study the fine and gross motor activities of crewmen in performing operations with and without the use of restraints.

Study crewmen performing tasks which require visual, tactile, or auditory feedback, or combinations of feedbacks.

Study intravehicular (IVA) and extravehicular (EVA) activities.

Study repeated activities performed early, midway, and late in the missions which will show adaptation to the zero-gravity environment.

16 MM CAMERA WITH QUICK UNIVERSAL MOUNT

PORTABLE HIGH INTENSITY PHOTOGRAPHIC LAMP

TYPICAL ACTIVITY TO BE PHOTOGRAPHED

FIGURE 186.—Experiment M151, Time and motion study of astronaut in various activities.

Instrumentation:

Recording of this experiment will be achieved with the 16 mm movie camera, and with a portable high intensity photographic lamp, both of which will be used also for other purposes. In addition, verbal information by the astronauts about their experiences in zero-gravity operations will be tape-recorded for later evaluation.

Similar recordings have been made on the ground during crew training.

M171, Metabolic Activity (Fig. 187)

Principal Investigator:
Edward L. Michel
NASA-Lyndon B. Johnson Space Center
Houston, Texas
Co-Investigator:
Dr. John A. Rummel
NASA-Lyndon B. Johnson Space Center
Houston, Texas
Development Center:
JSC; Integration Center: MSFC

Objective:

Determine man's metabolic effectiveness and its possible change while he is doing work in a zero-gravity environment. Obtain information on his physiological capabilities and limitations. Provide data useful in the design of future spacecraft and work programs.

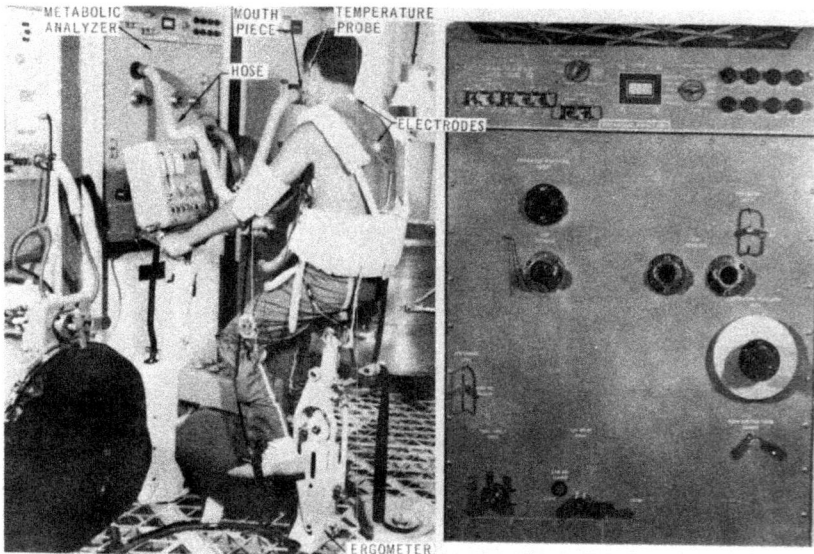

Figure 187.—Experiment M171, Study of metabolic activities of astronaut while exercising on Ergometer.

Physiological responses to physical activity will be deduced by analyzing inhaled and exhaled air, pulse rate, blood pressure, and other selected variables of a crew member performing controlled amounts of physical work with a bicycle ergometer.

Evaluate the bicycle ergometer as an exerciser on long-duration missions.

Collect data on ergometer work rate, ergometer RPM, oxygen uptake, carbon dioxide output, minute volume,[1] vital capacity, respiratory quotient, heart rate, blood pressure, vectorcardiogram, body weight, body temperature, and Skylab environmental parameters.

Instrumentation:

Main component of this experiment is an ergometer (wheelless exercising bicycle) whose pedal wheel friction is controlled by a cardiotachometer in such a way that a preselected heart rate of the crew member remains constant. It can also be controlled for a constant, preselected workload. Equipment further includes a respiratory gas analyzer, a blood pressure measuring system, a body temperature measuring system, and a vectorcardiogram system (see Experiment M093). A metabolic analyzer, containing a spirometer and a mass spectrometer, will measure oxygen uptake, carbon dioxide output, and minute volume.

Each crew member will perform this experiment five times during the 28-day mission, and eight times during the 56-day missions.

M172, Body Mass Measurement (Fig. 188)

Principal Investigator:
William E. Thornton, M.D.
NASA-Lyndon B. Johnson Space Center
Houston, Texas

Development Center:
JSC; Integration Center: MSFC

Objective:

Determine the body mass of each crew member; observe changes in body masses during flight. Demonstrate the proper functioning and assess the utility of the Body Mass Measuring System in daily use.

Knowledge of exact body mass variations throughout the flight will greatly help in the correlation of other medical data obtained during flight.

Instrumentation:

Mass measurements under zero-gravity conditions can be achieved by application of Newton's second law (force equals mass times acceleration). The force is provided by a spring; the mass is attached to a platform suspended as a pendulum by four parallel flexing strips. An optical pickup and an electric timer measure the period of the spring-loaded pendulum. In order to achieve a measurement, the astronaut sits in a

---

[1] Breathing volume per minute.

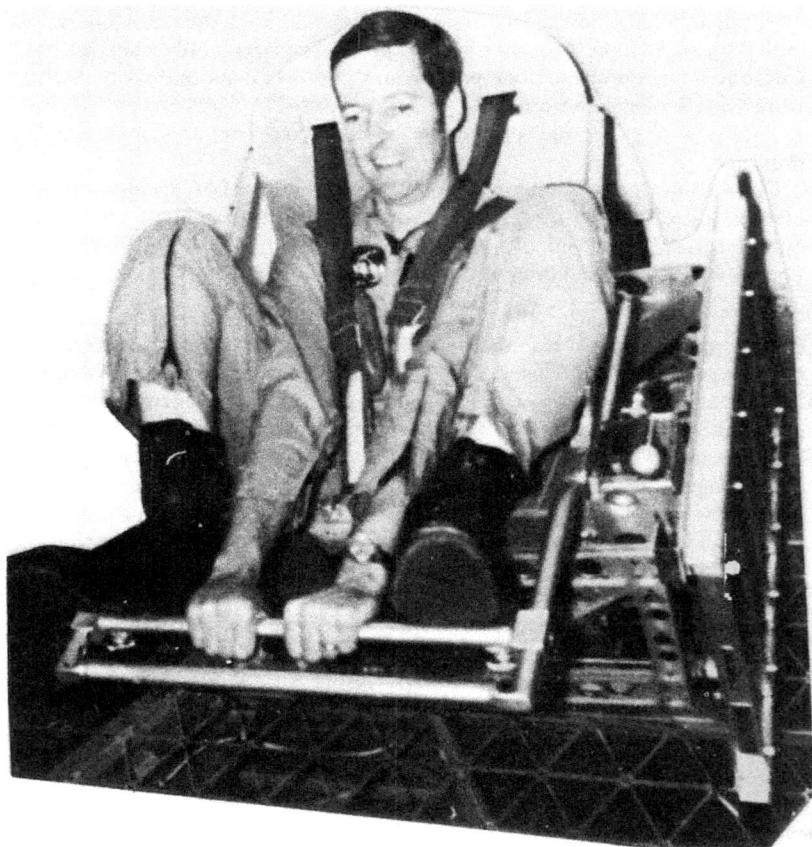

FIGURE 188.—Experiment M172, Determination of body mass.

compact posture on the platform, cocks the spring, and releases the platform which will oscillate with a sinusoidal motion. Time signals are converted to provide a readout directly in kilograms. (See Experiment M074). The Body Mass Measurement Device has been calibrated before launch with known masses up to 100 kg (220 lbs).

S015, Effects of Zero Gravity on Single Human Cells (Fig. 189)

Principal Investigator:
  Philip O. Montgomery, M.D.
  University of Texas
  Southwestern Medical School
  Dallas, Texas
Co-Investigator:
  Dr. J. Paul
  University of Texas

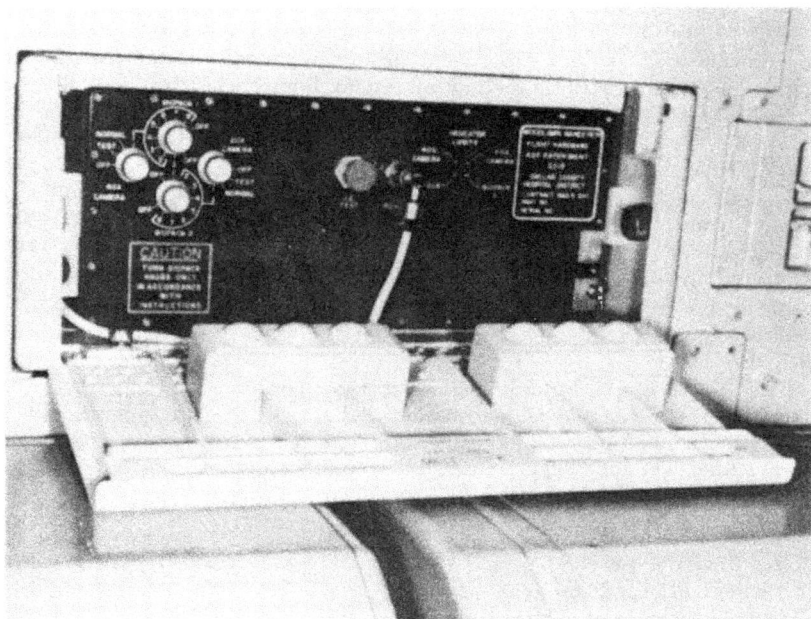

FIGURE 189.—Experiment S015, Study of zero gravity effects on human cells.

Southwestern Medical School
Dallas, Texas
Development Center:
    JSC; Integration Center: JSC

Objective:

Observe the functioning of living human cells in a tissue culture while they are weightless. Determine the effects of gravity on individual cells by making complete surveys of cellular structures and biochemical functions.

Similar observations have been made on human cells under the Earth's gravity and under high acceleration forces. Cells will be observed during flight with time-lapse microscope photography. Two separate cell cultures will be examined for four-day and ten-day periods. After several biochemical experiments have been performed, the cells will be preserved and returned to Earth for further biochemical testing. Their DNA, RNA, and lipide [1] content and their enzyme activity will be determined.

Control groups of human cells will be subjected to similar tests under Earth gravity and in high-acceleration fields.

---

[1] DNA=Dioxyribonucleic acid
RNA=Ribonucleic acid
Lipide
These substances are basic components of live cells (see glossary).

Instrumentation:

Instruments used for this experiment include a microscope-camera assembly and a growth curve module subsystem, both enclosed in a single hermetically sealed package. Two phase-contrast microscopes with magnifications of 20 and 40, each focused on its own specimen chamber, will provide images to the two 16 mm time-lapse cameras. The cameras will be cycled automatically by a built-in timing mechanism; each will run at 5 frames per minute for 40 minutes twice per day for the entire 28-day mission.

The specimen chambers will provide temperature-controlled environments for the cell cultures. Each chamber will have its own independent media exchange assembly to provide fresh nutrients to the cultures twice each day.

The Growth Curve Module consists of two operationally independent assemblies, each capable of maintaining living cells in nine chambers. At preprogrammed times, a fixative will be injected into eight chambers, one at a time. The cells will be returned for postflight analysis.

S071, Circadian Rhythm, Pocket Mice (Figs. 190, 191)

Principal Investigator:
Dr. Robert G. Lindberg
Northrop Corporation Laboratories
Hawthorne, California
Development Center:
Ames Research Center; Integration Center: JSC

Objective:

Determine whether the daily physiological rhythms of a mammal (pocket mouse, Fig. 140) are changed by a zero-gravity environment. Circadian rhythms (24-hour physiological wake-and-sleep cycles) of animals and man are suspected to be influenced to some extent by gravitational forces. Should an influence be discovered, this would be an indication that biorhythms at least of animals are timed and controlled by factors which include gravity.

Changed or affected rhythms alter the basic control of metabolism. It is important that normal biological rhythms in man be maintained for his well-being and effectiveness during space flight. If normal physiological rhythms are found to continue in the test animals, the conclusion can be drawn that space flight does not impose biorhythm restriction and that man can work in space without a degradation of his performance due to biorhythm disturbance.

Instrumentations:

The experiment includes six pocket mice placed in a completely dark cage at 15° C (60° F) temperature, relative humidity of 60° percent, and an atmospheric pressure equal to sea level pressure.

Three weeks before the mission, the mice will be placed in the cage. Body temperature and activity level will be automatically monitored to establish the natural period, phase, and stability of the rhythms under

FIGURE 190.—Experiment S071, Observation of pocket mice under zero gravity. Experiment S072, Observation of vinegar gnats (dropsophila) under zero gravity.

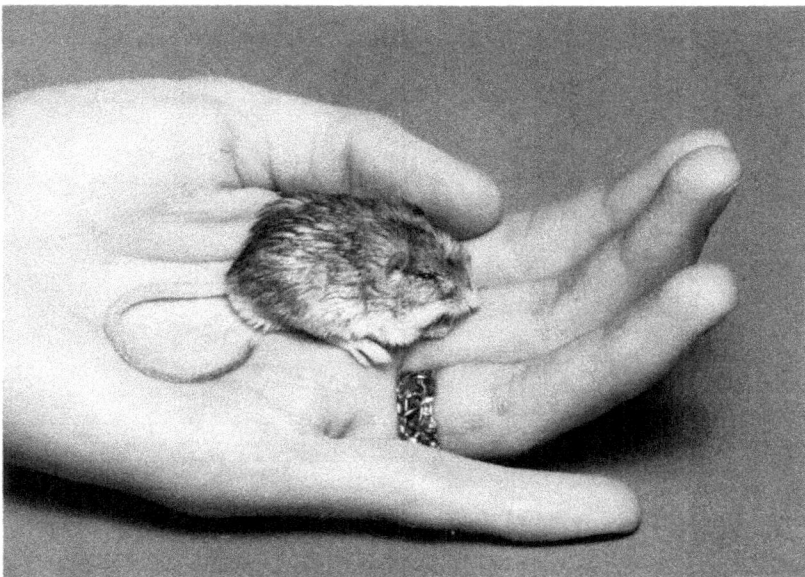

FIGURE 191.—Pocket mouse as used in Experiment S071.

183

Earth gravity conditions. The cage will be installed in the Service Module shortly before launch. The same measurements will be made during the flight. Data will be automatically recorded and telemetered to Earth for interpretation.

S072, Circadian Rhythm, Vinegar Gnats

Principal Investigator:
  Dr. Colin S. Pittendrigh
  Stanford University
  Stanford, California
Development Center:
  Ames Research Center; Integration Center: JSC

Objective:

Determine whether the daily emerging cycle of the vinegar gnat (drosophila) from the pupa to the fly is the same under weightlessness and on Earth.

Extensive experiments have shown that even though gnats in the pupa stage develop at different rates depending on temperature, the adult gnats will not emerge from their pupae until some kind of internal signal is given. This triggering signal is somehow timed to occur at an exactly fixed time delay after a flash of light. The signal occurs at the same constant time interval after the flash, independent of the temperature.

The experiment will measure the emergence times of four groups at 20° C (68° F) to find out whether space flight conditions change the mechanism which keeps the rhythm constant despite changes in temperatures.

Each group of pupae is divided into two subgroups. Flashes of light will initiate the emergence of the two subgroups at two different times. If the delayed subgroup shows the same rhythm of emergence response as the earlier group, it is probable that no external factor contributes to the rhythm and that the rhythms are internally synchronized with the light flashes.

This experiment is conducted in conjunction with the pocket mice experiment (S071). If rhythms of both experiments are disrupted or altered during space flight, it can be assumed that space flight disrupts or alters the common basic rhythm mechanisms, and that man's biological rhythm mechanism is probably also affected by zero gravity.

Instrumentation:

This experiment is designed for automatic operation. After the initial white light flash has occurred for initiation of the emergence cycle, a dim red light will be turned on every 10 minutes, and 180 pupae photocells will be scanned electronically. After a gnat has emerged, its pupa shell will be transparent, and the pupa photocell will respond to the red light.

Emergence data will be stored and later telemetered to Earth.

Experiment Support System

This facility (Fig. 192), located in the Orbital Workshop, will provide normal power, special regulated power, controls and displays, data management, recording facilities, programmed time signals, pressurized gas, and calibration commands for the biomedical experiments M092, M093, M131, and M171. Specific subsystems will support blood pressure measurements, leg volume measurements, and vectorcardiogram measurements.

The Experiment Support System was developed and integrated by MSFC.

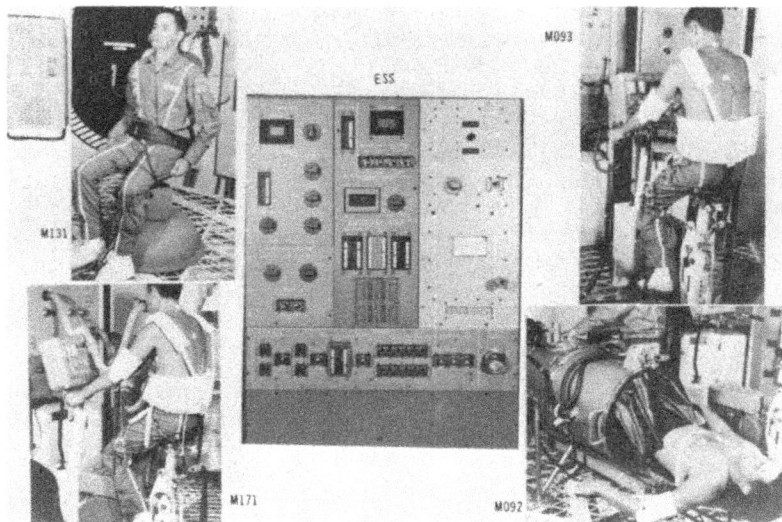

FIGURE 192.—Experiment Support System.

4. SPACE TECHNOLOGY PROJECTS

Skylab provides a unique opportunity to develop a better understanding of the effects of the space environment on materials and operational devices. Skylab can also be used as a platform to measure the environment within and outside of a spacecraft, and man's influence upon this environment. A number of Skylab experiments will provide data about engineering and technological operations in the space environment; these data will be important in the development of future Earth orbital space stations and spacecraft. Thus, through the Skylab program, it will be possible to learn more about how man performs in space, what tools he needs to accomplish his tasks, what his influence is on the space environment, and how materials can be processed in space.

The space technology experiments on Skylab can be divided into the following categories; they will be grouped in this order in the descriptions that follow.

Material Science and Manufacturing in Space

M479 Zero-gravity Flammability
M512 Materials Processing in Space
M551 Metals Melting
M552 Sphere Forming
M553 Exothermic Brazing
M555 Gallium Arsenide Crystal Growth
M518 Multipurpose Electric Furnace System
M556 Vapor Growth of II–VI Compounds
M557 Immisciple Alloy Compositions
M558 Radioactive Tracer Diffusion
M559 Microsegregation in Germanium
M560 Growth of Spherical Crystals
M561 Whisker-Reinforced Composites
M562 Indium-Antimonide Crystals
M563 Mixed III–V Crystal Growth
M564 Halide Eutectics
M565 Silver Grids Melted in Space
M566 Copper-Aluminum Eutectic

Zero-Gravity Systems Studies

M487 Habitability/Crew Quarters
M509 Astronaut Maneuvering Equipment
M516 Crew Activities and Maintenance Study
T002 Manual Navigation Sightings
T013 Crew/Vehicle Disturbance
T020 Foot-Controlled Maneuvering Unit

Spacecraft Environment

D008 Radiation in Spacecraft
D024 Thermal Control Coatings
M415 Thermal Control Coatings
T003 Inflight Aerosol Analysis
T025 Coronagraph Contamination Measurements
T027 ATM Contamination Measurements

a. Astronaut Tools and Equipment

In addition to specific experiments, there is a considerable array of tools, miscellaneous supplies, and support equipment for the crewmen in the Skylab (Fig. 193), including tool kits, repair kits, restraints, supplies, a film vault, photographic equipment, and TV cameras. Two tool kits are installed and retained in standard stowage lockers. The tools provided include standard ranges and sizes of sockets, open end/box wrenches, screwdrivers, and screwdriver bits. Also included are a vise, a speeder handle, a spin-type handle, a ratchet handle, a pin straightener, and other common handtools.

A repair kit is also installed in a standard stowage locker. This kit contains the necessary types and sizes of blister patches to repair structural leaks. Additional items provided include flat patches, Teflon tape, sealant putty, Velcro fasteners, restraints, scissors, and tape for repairing air duct damage.

Another support kit includes tension and compression scales, a steel measuring tape, a sound level meter, a frequency analyzer, two surface temperature digital thermometers, three ambient thermometers, and an air velocity measuring instrument.

Cameras using both film and television are provided. To support the film cameras, there is a film vault to provide protection from radiation (Fig. 194). Photographic lights, power and signal cables, versatile mobile restraints, and convenience outlets are provided. The film cameras available for use inside the Skylab include a 16 mm Data Acquisition Camera (frame rates 2, 4, 6, 12 and 24 frames per second with shutter speeds from 1/60 to 1/1,000 of a second), 35 mm, and 70 mm still cameras. Speed, resolution, and coverage depend upon film and lens selected.

b. Materials Science and Manufacturing in Space

The zero-gravity condition existing in Skylab makes it possible to perform operations in materials processing that would be impossible or extremely difficult on Earth. Melting and mixing without the contaminating effects of containers, the suppression of convection and buoyance in liquids and molten

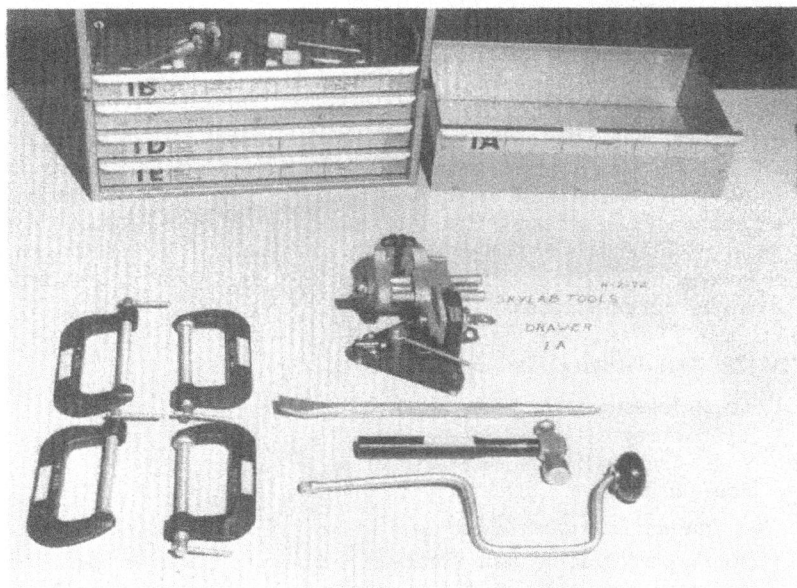

FIGURE 193.—Tool Kit and typical tools for astronauts.

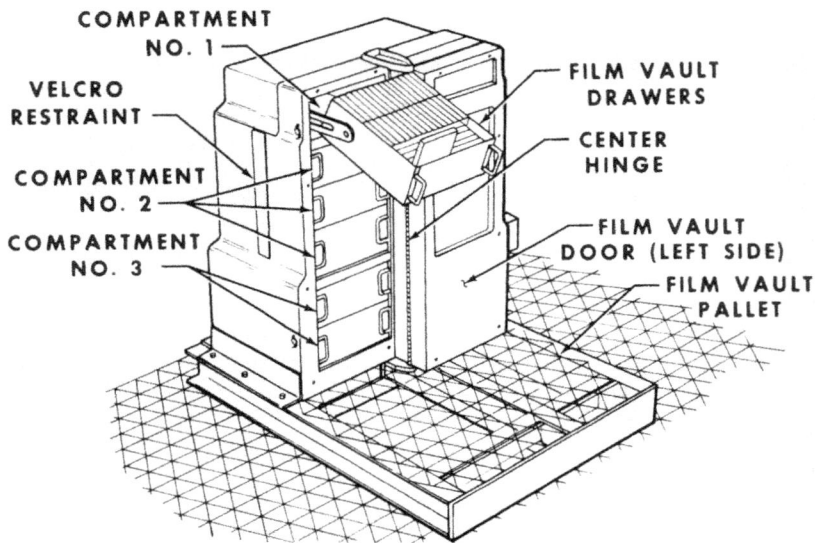

FIGURE 194.—Film Vault for storage of photographic films.

material, control of voids, and the ability to use electrostatic and magnetic forces otherwise masked by gravitation open the way to new knowledge of material properties and processes and ultimately to the production of valuable new materials for use on Earth. These potential products range from composite structural materials with specialized physical properties to large and highly perfect crystals with valuable electrical and optical properties. In addition, it will be possible to evaluate the feasibility of using electron beam and thermo-welding under zero-gravity conditions.

Practical experience with principles and problems of development and integration which has been gained in developing the Skylab materials processing facility has already proved very valuable in the concept planning of an improved and enlarged facility for incorporation into the Space Shuttle Program. Final facility design for Shuttle missions will be based upon the evaluation of the Skylab program results, and after detailed user requirements have been specified.

M479, Zero-Gravity Flammability

Principal Investigator:
J. H. Kimzey
NASA-Lyndon B. Johnson Space Center
Houston, Texas

Objective and Instrumentation:

Obtain photographic data of various combustible materials ignited under controlled conditions in a zero-gravity environment to determine the extent of flame propagation and flashover to adjacent materials, rates of surface and bulk flame propagation under zero-convection, self-extinguishment characteristics, and extinguishment characteristics by vacuum or water spray. The combustion chamber and controls for this experiment are provided by Experiment M512 (Fig. 195). The combustion chamber is stainless steel with a low emissivity interior (Fig. 196). A large opening on one end permits installation of igniter-fuel assemblies. Chamber connections are provided for venting either to the vacuum of space to exhaust smoke and products of combustion, or venting to the vehicle interior to equalize pressure and to permit opening. Interior work lights are provided as well as means to remove solid ash particles by a vacuum cleaner equipped with a filter trap. Provisions are also made to spray water to evaluate extinguishment by that means. The igniter-fuel assemblies are housed in a separate container which serves both as a protection to the assemblies and as a place to dispose of used assemblies after testing. Data will be recorded on motion picture film to be returned to Earth for analysis. The astronaut will provide voice comments while performing the experiment.

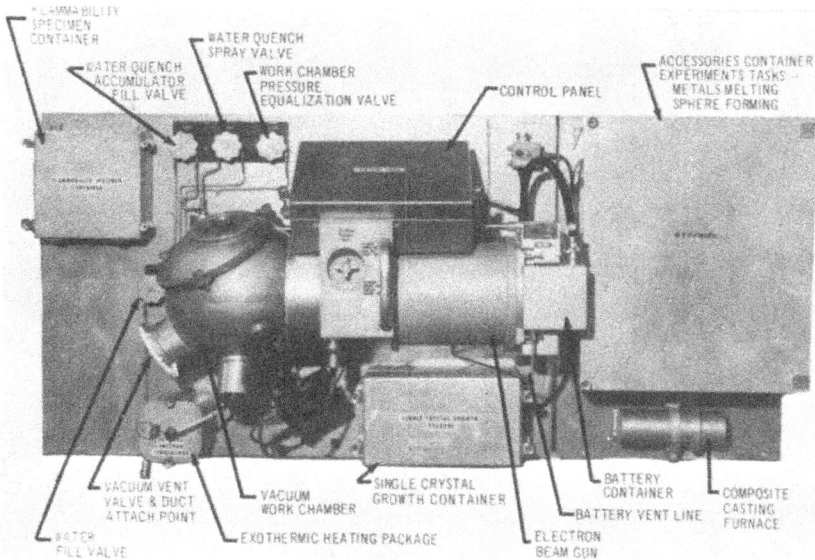

FIGURE 195.—Experiment M512, Materials processing in space.

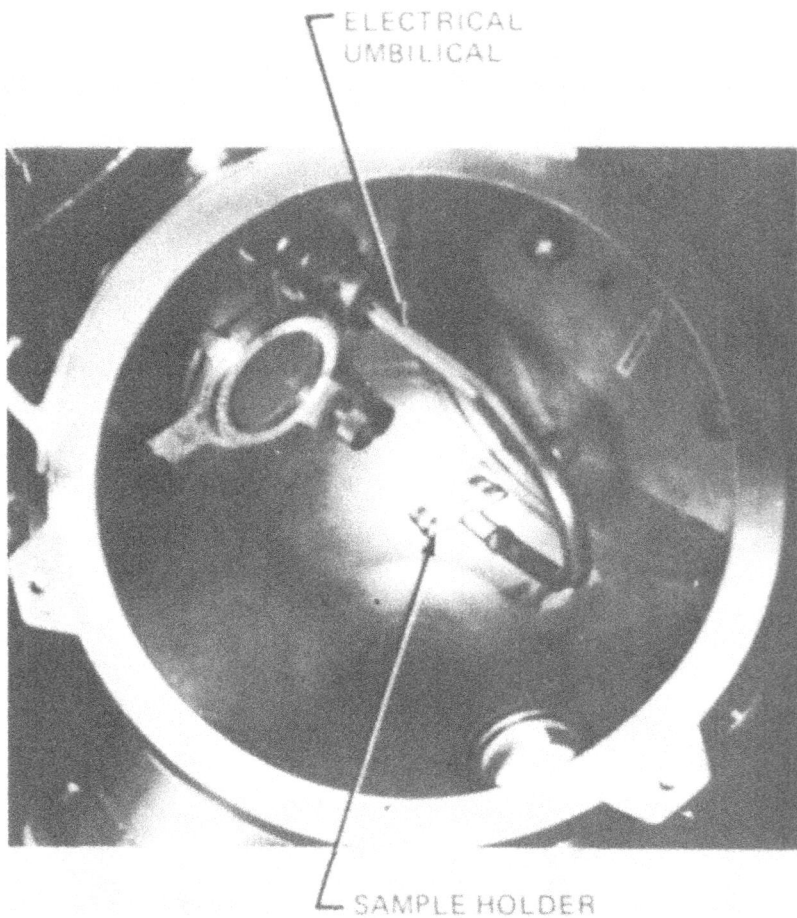

ELECTRICAL
UMBILICAL

SAMPLE HOLDER

FIGURE 196.—Experiment M479, Flammability under zero gravity.

M512, Materials Processing Facility (Fig. 197)

Principal Investigator:
  P. G. Parks
  NASA-Marshall Space Flight Center
  Huntsville, Alabama

Objective and Instrumentation:

Perform fundamental research on the effects of zero-gravity on molten metal processing. All the tasks belonging to this experiment involve the melting of materials by the application of heat. On Earth, the density differences caused by temperature differences result, under the influence of gravity, in convection. For many purposes, this is a hindrance on Earth; however, in zero gravity there will be no convection due to temperature differences.

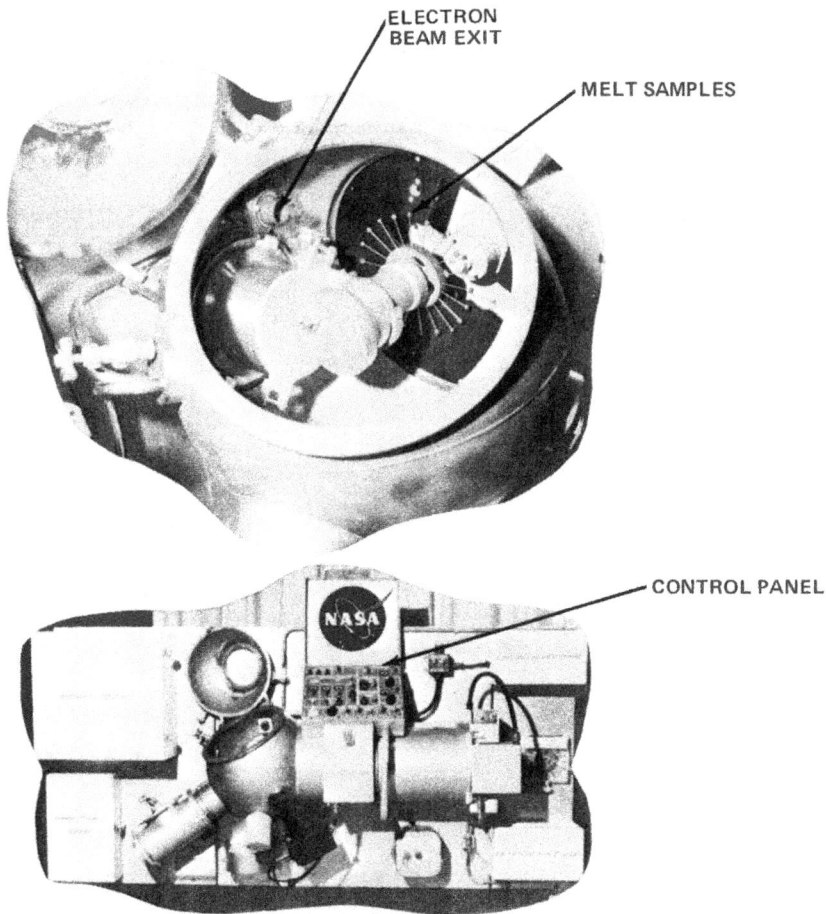

FIGURE 197.—Experiment M512, Control panel and electron beam gun.

Another effect of gravity that will be avoided in space processing is the separation of different density materials in the preparation of composites. Certain materials of superior characteristics could be produced if a uniform or other preferred mixture of substances of different densities could be attained. On Earth, fibers or particles embedded in a melt will either float or settle if their density differs from that of the matrix, but in space this will not occur.

The place where materials processing and a number of other experiments on manufacturing in space will be carried out is the M512 facility. This facility, mounted in the MDA, consists of a vacuum work chamber with associated mechanical and electrical controls, an electron beam subsystem (Fig. 198), and a control and display panel. The vacuum chamber is a 40-cm (16-in) sphere with a hinged hatch for access. It is connected to the space environment by a 10-cm (4-in) diameter line containing two gate valves. The electron beam subsystem is mounted to the chamber so that the beam traverses the sphere along a diameter parallel to the plane of the hatch closure. The chamber wall contains a cylindrical well accommodating the small electric furnaces used for the M555 experiment. A receptacle above the well provides power and instrumentation lead connections to the control panel. Auxiliary provisions include ports for a floodlight and the 16 mm data acquisition camera, a bleed line, a repressurization line, and a port for a vacuum cleaner to remove debris from the chamber. A subsystem is also provided for spraying water into the chamber during some runs of the M479 experiment.

The electron beam operates nominally at 20 kilovolts and 80 milliamperes. Focusing and deflection coils can be operated from the control panel to adjust the size and position of the beam impingement spot on the experiment samples.

The control panel contains controls and displays for all experiments to be performed in the facility, including a pressure gauge for the vacuum chamber, voltage and current meters for the electron beam, and a thermocouple temperature indicator.

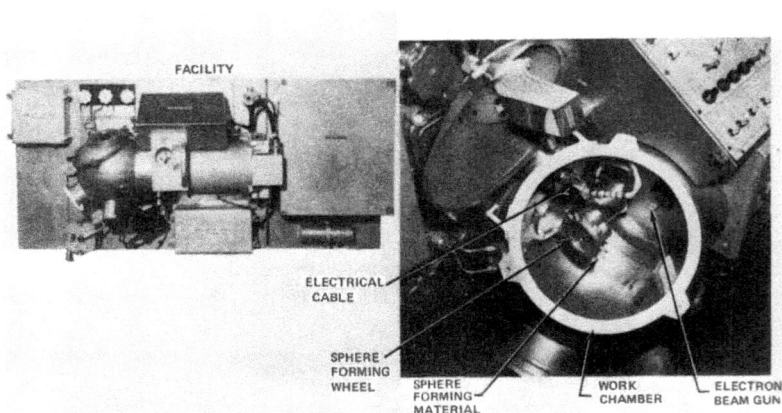

FIGURE 198.—Experiment M512, Electron beam gun.

Functions of the crew will include installing specialized apparatus and samples for each experiment in the chamber, operation of the experiments from the control panel, observation of some experiments through a viewport in the chamber hatch, data recording, and disassembly of each experiment after it is performed.

The control panel of the M512 facility permits switch settings and dial readings (Fig. 197). All data taken from the control panel must be recorded in writing or on voice recorder tape by the crew.

Data from the experiments will comprise the samples, those parts of the apparatus that are to be returned, motion picture records of the two electron beam experiments and M479, and comments by the operating crewmen. The returned samples will be studied in comparison with control samples produced on Earth.

Four tasks are contained in this experiment:

M551, Metals Melting—Examine the molten metal flow characteristics of various metal alloys. Metal samples will be melted by electron beam and photographed. Specimens will be analyzed after return. Principal Investigator: R. M. Poorman, Marshall Space Flight Center, Huntsville, Alabama.

M552, Exothermic Brazing—Develop a stainless steel tube joining technique for assembly and repair in space; evaluate the flow and capillary action of molten braze material, using exothermic material as heat source; and demonstrate the feasibility of exothermic reaction brazing in space. Principal Investigator: J. Williams, Marshall Space Flight Center, Huntsville, Alabama.

M553, Sphere Forming (Fig. 198)—Produce spherical shapes of about 6 mm (.236 in) diameter from molten specimens of metals under the forces of surface tension by taking advantage of the virtual absence of gravity. Spheres will be evaluated on Earth. Principal Investigator: E. A. Hasemeyer, Marshall Space Flight Center, Huntsville, Alabama.

M555, GaAs Crystal Growth—Grow single crystals of gallium arsenide of exceptionally high chemical purity and crystalline perfection. GaAs will dissolve in liquid gallium metal at the hot end (750° C or 1382° F) of a quartz tube; crystals will form on seed crystals at the cold end (550° C or 1022° F) of the tube. The tube will be opened for crystal analysis after return to Earth. Principal Investigator: Dr. M. Rubenstein, Westinghouse Electric Corporation, Pittsburgh, Pennsylvania.

M518, Multipurpose Electric Furnace System (Fig. 199)

Project Engineer:
Arthur Boese
NASA-Marshall Space Flight Center
Huntsville, Alabama

Objective and Instrumentation:

Provide a means for experimentation in solidification, crystal growth, composite structures, alloy structural characteristics, and other thermal proc-

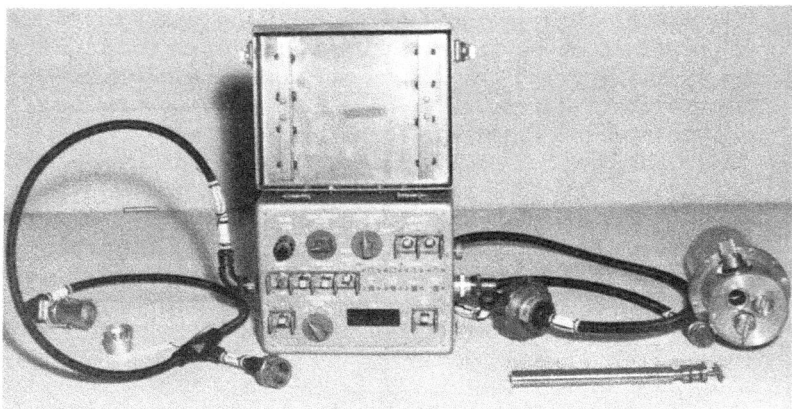

FIGURE 199.—Multipurpose Electric Furnace and control panel.

esses involving changes in materials under conditions of weightlessness. The system consists of three main units, the Multipurpose Furnace, the Control Package, and 33 cartridges (11 experiment sets). The Furnace has three specimen cavities so that three material samples (cartridges) can be processed in a single run. The furnace is designed to provide three different temperature zones along the length of each sample cavity, as follows:

(1) A constant temperature hot zone at the end of the sample cavity where temperatures up to 1000° C (1832° F) can be reached,

(2) A gradient zone next to the hot zone where temperature gradients ranging from 20° C (36° F) per centimeter to 200° C (360° F) per centimeter can be established in the samples, and

(3) A cool zone in which heat conducted along the samples will be radiated to a conducting path that carries the heat out of the system.

The control package, providing active control of the furnace temperature, can be set to any specified temperature within the furnace's capability (0° to 1000°C or 32°F to 1832°F) by the astronaut operating the system. Two timing circuits in the controller will enable the astronaut to program the soak time spent at the set temperature and the cooling rate of the furnace following the end of the soak period. The cartridge encapsulates the sample material, and the actual temperature distribution applied to the sample will be controlled by the thermal design of the cartridge. Once the specimens are installed in the furnace and the system is activated, the Multipurpose Electric Furnace System will operate automatically except for complete system shutdown.

The eleven experiments planned for the M518 system and their objectives are:

M556—Vapor Growth of II–VI Compounds [1]—Determine the degree

---

[1] Compounds formed of elements of the II and the VI group of the Periodic System of Elements. Such compounds are mostly semiconductors.

of improvement that can be obtained in the perfection and chemical homogeneity of crystals grown by vapor transport under weightless conditions. Principal Investigator: Dr. H. Wiedemeier, Rensselaer Polytechnic Institute, Troy, New York.

M557—Immiscible Alloy Compositions—Determine the effects of near zero-gravity on the processing of material compositions which segregate in the melt on Earth because of density differences. Principal Investigator: J. Reger, Thompson Ramo Wooldridge, Redondo Beach, California.

M558—Radioactive Tracer Diffusion—Measure self-diffusion and impurity diffusion effects in liquid metal under zero gravity, and characterize the disturbing effects, if any, as a consequence of spacecraft acceleration. Principal Investigator: Dr. T. Ukanwa, NASA-Marshall Space Flight Center, Huntsville, Alabama.

M559—Microsegregation in Germanium—Determine the degree of microsegregation of doping impurities in germanium caused by convectionless directional solidification under conditions of weightlessness. Principal Investigator: Dr. F. Padovani, Texas Instruments, Dallas, Texas.

M560—Growth of Spherical Crystals—Grow doped germanium crystals of high chemical homogeneity and structural perfection, and compare their physical properties with theoretical values for ideal crystals. Principal Investigator: Dr. H. Walter, University of Alabama in Huntsville, Huntsville, Alabama.

M561—Whisker-Reinforced Composites—Produce void-free samples of silver and aluminum reinforced with oriented silicon carbide whiskers. Principal Investigator: Dr. T. Kawada, National Research Institute for Metals, Tokyo, Japan.

M562—Indium Antimonide Crystals—Produce doped semiconductor crystals of high chemical homogeneity and structural perfection, and evaluate the influence of weightlessness in attaining these properties. Principal Investigator: Dr. H. Gatos, Massachusetts Institute of Technology, Cambridge, Massachusetts.

M563—Mixed III–V Crystal Growth [1]—Determine how weightlessness affects directionality of binary semiconductor alloys.

If single crystals are obtained, determine how their semiconducting properties depend on alloy composition. Principal Investigator: Dr. W. Wilcox, University of Southern California, Los Angeles, California.

M564—Halide Eutectics—Produce samples of the fiberlike sodium chloride-sodium fluoride eutectic, and measure its physical properties. In particular, optical parameters of the space-produced material will be of interest. Principal Investigator: Dr. A. Yue, University of California at Los Angeles, Los Angeles, California.

M565—Silver Grids Melted in Space—Determine how pore sizes and pore shapes change in grids of fine silver wires when they are melted and resolidified in space. Principal Investigator: Dr. A. Deruythere, Catholic University of Leuven, Heverlee, Belgium.

---

[1] Compounds formed of elements of the III and V group of the Periodic System of Elements.

M566—Copper-Aluminum Eutectic—Determine the effects of weightlessness on the formation of lamellar structures in eutectic alloys when directionally solidified. Principal Investigator: E. Hasemeyer, NASA-Marshall Space Flight Center, Huntsville, Alabama.

c. Zero-Gravity Systems Studies

The zero-gravity systems studies experiments were selected to gain engineering data on man's capabilities in the weightless environment and to evaluate instruments and systems designed to increase the crew's ability to move and work in such an environment. These investigations support the broad Skylab program objectives of developing and advancing man's capability to live and perform useful work in space. Evaluations will be made of the crew's ability to perform zero-gravity tasks requiring both gross and delicate manipulations. A series of measurements will be taken to determine the effects of mission duration and prolonged weightlessness on the crew's ability to repeat these same tasks in a skillful and timely manner. Accurate measurements will be made of vehicle disturbances caused by crew activities. These results are essential for the definition of allowable crew motions during the conduct of experiments requiring low levels of acceleration (10^{-4} to 10^{-5} g) or pointing with extreme accuracy and for the design of attitude control systems for future space missions. Other experiments will evaluate operating concepts for astronaut maneuvering devices. The data obtained will be applied to future programs for extra-vehicular activities such as assembling large space structures, inspection and maintenance, repair, refurbishment, rescue, and data retrieval. Several astronaut maneuvering units and labor saving aids will be evaluated during the mission.

M487, Habitability/Crew Quarters

Principal Investigator:
C. C. Johnson
NASA-Lyndon B. Johnson Space Center
Houston, Texas

Objective and Instrumentation:

Evaluate the features of Skylab's living quarters, crew provisions, and support facilities as they affect the crew's comfort, safety, and operating efficiency. Equipment, procedures, and habitat design concepts derived from experience on Earth and from previous short-duration orbital flights, may require modification. This evaluation is a multidisciplinary set of systematic observations, and it serves as a test and validation of design concepts and technical features. The following aspects of system design and operation will be studied: physical environment (temperature, humidity, light, noise); architecture (volume and layout of working and living areas); mobility aids and personal restraints (translation, worksite support, sleep stations); food and water (storage, preparation, quality); personal garments (comfort, durability, design); personal hygiene (cleansing, grooming, collection and disposal of body waste); housekeeping (habitat cleansing, waste control and disposal) ; off-duty activities (exercise facilities, individual and group recreation, privacy features) ; and communications.

Instruments used in this study include a portable surface temperature digital thermometer, sound level meter, frequency analyzer, air velocity meter, measuring tape, and ambient thermometers. Motion picture cameras, lights, and tape recorders will be available from other experiments. Data will be recorded in the form of motion picture films and voice tape comments. The data will be evaluated on Earth after they are returned.

M509, Astronaut Maneuvering Equipment (Fig. 200)

Principal Investigator:
Major C. E. Whitsett, Jr., USAF
Air Force Space and Missile Systems Organzation
Los Angeles, California

Objectives and Instrumentation:

Conduct an in-orbit verification of the utility of various maneuvering techniques to assist astronauts in performing tasks which are representative of future extravehicular activity (EVA) requirements.

The concept of powered astronaut maneuvering is fundamental to the development of an effective EVA capability which, in turn, is considered

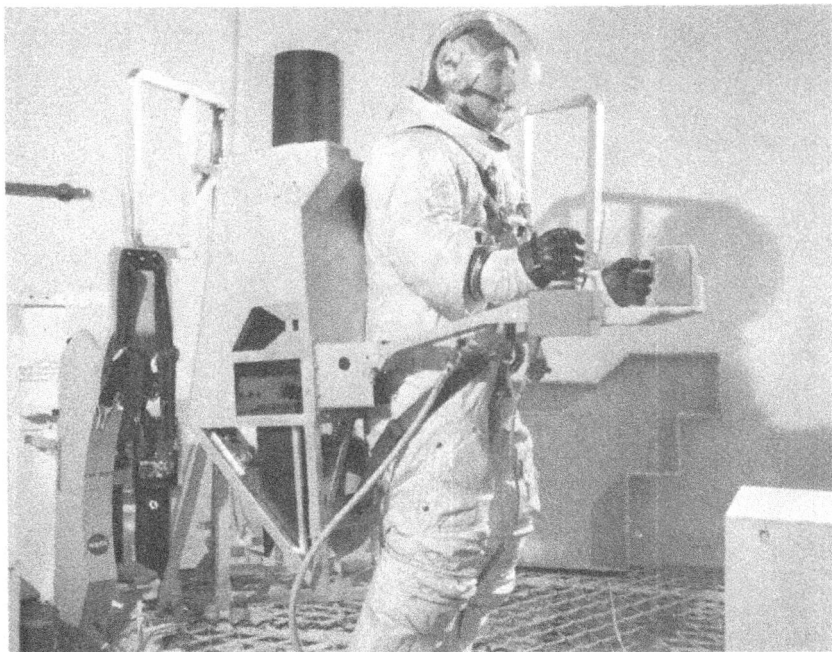

FIGURE 200.—Experiment M509, Space maneuvering system for use during extravehicular activities.

to be a routine supporting element in future manned space flight. EVA is expected to play a major role in such areas as space rescue, inspection and repair of parent and satellite spacecraft, personnel and cargo transport, and space structure erection. The addition of maneuvering aids to such EVA tasks is expected to reduce crew fatigue and stress, cut time requirements, offset pressure suit mobility limitations, and facilitate attitude orientation and stabilization.

The astronaut maneuvering equipment of Experiment M509 consists of two jet-powered aids for maneuvering in a zero-gravity space environment. The first is a back-mounted, hand-controlled Automatically Stabilized Maneuvering Unit (ASMU); the second a Hand-Held Maneuvering Unit (HHMU). A backpack, serving both units, contains a rechargeable or replaceable high pressure nitrogen propellant tank. It will be worn for the ASMU and for the HHMU (Fig. 201). The electrical systems within the backpack are powered by a rechargeable or replaceable battery. The astronaut dons the backpack over either a pressurized space suit or flight coveralls, using a quick release harness similar to that used for parachutes.

The Automatically Stabilized Maneuvering Unit is maneuvered in six degrees of freedom (X, Y, and Z axis translation, and pitch, yaw, and roll) by means of 14 fixed thrusters located in various positions on the backpack. Control of the thrusters is achieved by two hand-controllers mounted on arms extending from the backpack. The controllers are identical to

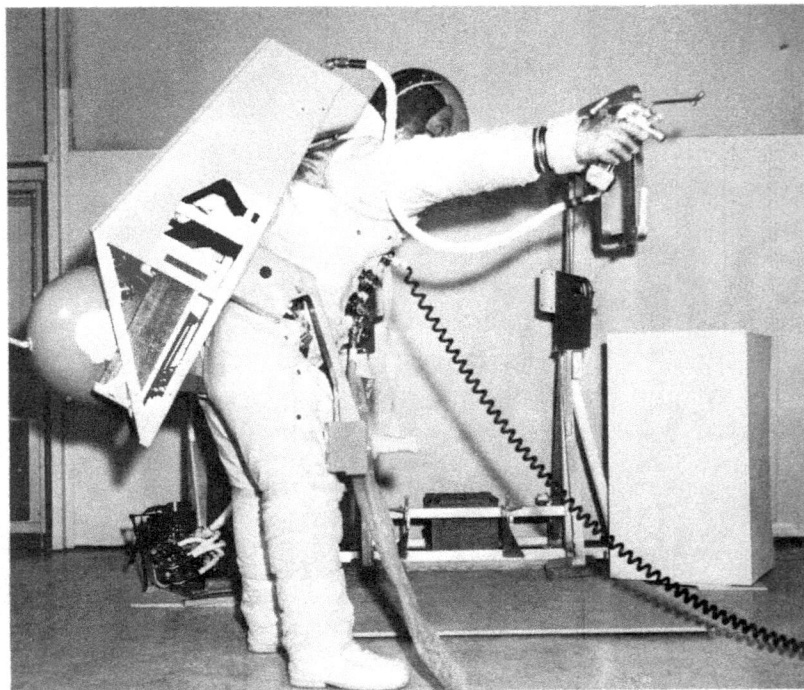

FIGURE 201.—Simulated use of hand-controlled maneuvering unit.

those used in the Apollo spacecraft. The Hand-Held Maneuvering Unit is a simple, small, lightweight, completely manual device similar to the one used in Gemini. It consists of a hand grip and controls for a pair of tractor (pull) thrusters and an opposing single pusher thruster; the assembly is connected to the ASMU propellant tank by a short hose. To orient and propel himself in any attitude or direction, the operator points the HHMU, aligns it so that the thrust vector passes approximately through his center of gravity, and triggers the tractor or pusher thrusters as indicated by his visual cues. Maneuvering with the ASMU and the HHMU on Skylab will be performed within the Orbital Workshop.

The ASMU is instrumented to record numerous engineering and biomedical data during the pressure-suited runs. These data will be sensed, collected, and telemetered from the free-flying ASMU to a receiver within the Orbital Workshop. Together with recorded voice commentary, the data will be telemetered from the OWS to ground stations. Additional experiment data will be provided by inflight television, postflight still and motion picture data, and logbook entries. It is expected that M509 will provide a wide range of valuable information on maneuvering unit handling qualities, operating techniques, consumable requirements, capabilities, and limitations.

M516, Crew Activities and Maintenance Study

Principal Investigator:
R. L. Bond
NASA-Lyndon B. Johnson Space Center
Houston, Texas

Objective:

To investigate crew performance in zero gravity, long-duration missions, primarily through observations of normal Skylab tasks. This experiment is oriented toward the design of future space equipment and work provisions rather than basic human performance. The experiment calls for systematic documentation of man's performance during prolonged weightless space flight, acquisition and evaluation of inflight maintenance data, and evaluation of data relative to design criteria for future manned missions. There is no requirement for any special equipment solely for this experiment. Data will be acquired for normal Skylab inflight activities. They will be recorded by film, voice tapes, logbook entries, TV transmissions, and telemetry.

T002, Manual Navigation Sightings (Fig. 202)

Principal Investigator:
Robert J. Randle
NASA-Ames Research Center
Moffett Field, California

Objective and Instrumentation:

Investigate the effects of the space flight environment (including long-mission time) on the navigator's ability to take space navigation measure-

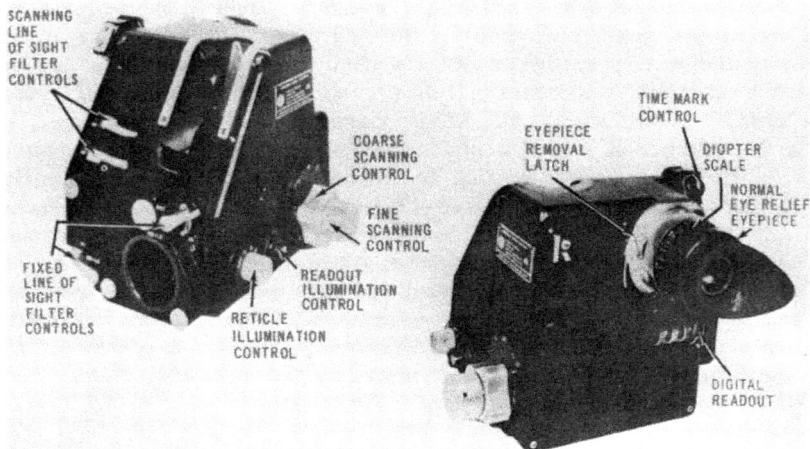

FIGURE 202.—Experiment T002, Manual navigation sighting instrument used to measure angles between two stars.

ments through a spacecraft window using hand-held instruments. Previous data obtained with the use of simulators, aircraft, and the Gemini spacecraft have already demonstrated that man, in a space environment, can make accurate navigation measurements using simple hand-held instruments. The intent of this experiment is to determine whether long-mission duration appreciably affects the capability of man to obtain accurate measurements. Further, the experiment will return data which will be generally indicative of the effect of long-duration space flight on man's capability to perform other precision tasks. The instrumentation for this experiment consists of a sextant and a stadimeter, both hand-held. The sextant, similar to an aviator's sextant, will be used to measure the angles between two stars and between single stars and the edge of the Moon. The stadimeter, also an optical device, determines spacecraft altitude directly by measuring the apparent difference in elevation angle between a portion of the Earth's horizon and its subtended chord. Data return will be in the form of logbook entries of the sextant and stadimeter readings. This will be supplemented by crew comments on the voice tape recorder.

T013, Crew/Vehicle Disturbance

Principal Investigator:
Bruce A. Conway
NASA-Langley Research Center
Hampton, Virginia

Objective and Instrumentation:
Determine the effect of the crew's activities within the spacecraft on the Skylab vehicle's pointing stability. Many Earth-pointing and astronomy experiments in future manned space programs will require pointing ac-

curacies of fractions of a second of arc. One of the most significant hindrances to achieving this accuracy will probably be the movement of the astronauts operating the spacecraft. Adequate design of the pointing control system for these future vehicles demands accurate knowledge of these effects. In this experiment, the forces exerted on the spacecraft by specific astronaut body and limb movements will be precisely measured. A limb motion sensor, attached to a suit, will measure the relative motions of the body, upper arm, lower arm, and upper and lower leg. The astronaut performing the experiment will be attached to a device that measures the force exerted on the Skylab structure by his activities. The limb movements and forces will be recorded on tape, and the activity will be photographed on motion picture film.

T020, Foot-Controlled Maneuvering Unit (Fig. 203)

Principal Investigator:
Donald E. Hewes
NASA-Langley Research Center
Hampton, Virginia

Objective and Instrumentation:

Evaluate an astronaut maneuvering device that does not require use of the astronaut's hands. The Foot-Controlled Maneuvering Unit (FCMU) is a research apparatus for examining the maneuvering dynamics of a cold gas jet-powered personal propulsion system in a zero-gravity space environment. The FCMU is propelled by high pressure nitrogen supplied from the M509 back-mounted tank. The operator, wearing either a pressurized space suit or flight coveralls, will control pitch, yaw, roll, and translation along his head-foot axis through a combination of toe and foot commands. Both the FCMU propellant tank and battery are rechargeable or replaceable units; they are shared with the astronaut maneuvering equipment of Experiment M509. Most of the data collection will be accomplished by two motion picture cameras, one mounted in the workshop dome and a battery-powered, forward-looking camera mounted within the FCMU frame. Additional data will be supplied by recorded voice commentary and logbook entries. It is expected that the information derived from this experiment will add valuable engineering inputs into future maneuvering unit design.

d. Spacecraft Environment

The Spacecraft Environment experiments on Skylab are concerned with measuring the radiation in the spacecraft, contamination around the spacecraft, and the effects of the spacecraft environment on thermal control coatings

The major source of radiation encountered by a spacecraft in Earth orbit is the South Atlantic Anomaly, a region where the Van Allen radiation belts are unusually close to Earth because of the unsymmetrical shape of the Earth's magnetic field. In addition, however, major solar flares can generate high energy protons and alpha particles. Also, there is continuous background radiation from cosmic ray sources. Measurements of the radiation

Figure 203.—Experiment T020, Foot-controlled maneuvering unit for extravehicular activities.

environment are needed to better predict the radiation dose received by man in Earth orbit.

Thermal control coatings with selected absorption and emission properties are applied to spacecraft surfaces to aid in maintaining the desired temperatures inside the spacecraft. Unfortunately, the environments to which these coatings are exposed before and during flight often alter their properties in a sense of making them less effective. Two Skylab experiments are designed to measure the effects of the space environment upon thermal control coatings.

Clouds of particles surrounding a spacecraft are termed contamination; this contamination may lead to deposits on optical surfaces, and it also may degrade visibility around the spacecraft. Particles originate from thruster firings, water and urine dumps, and outgassing of spacecraft surfaces. Two Skylab experiments will measure the contamination environment around Skylab.

D008, Radiation in Spacecraft (Fig. 204)

Principal Investigator:
Capt. Andrew D. Grimm, USAF
Kirtland AFB, New Mexico

Objective and Instrumentation:

Make radiation dose measurements in Earth orbit. These measurements are of value in assessing the quality of dosimetry instrumentation for use in space, in evaluating analytical procedures for predicting radiation doses in Earth orbit, and in studying the biological reaction of man to radiation. Instrumentation for this experiment consists of one portable tissue-equivalent dosimeter, one Linear Energy Transfer system comprised of two solid-state particle detectors which measure equivalent ranges in tissue of incident particles, and five passive dosimeters. The passive dosimeters and the Linear Energy Transfer system are placed in specific locations and remain there throughout the Skylab mission. The tissue-equivalent dosimeter will be moved to various locations in the spacecraft during measurement periods. Data from the tissue-equivalent dosimeter and the Linear Energy Transfer system will be telemetered to Earth; in addition, the logbook of events during measurement periods and the passive dosimeters will be returned to Earth for analysis.

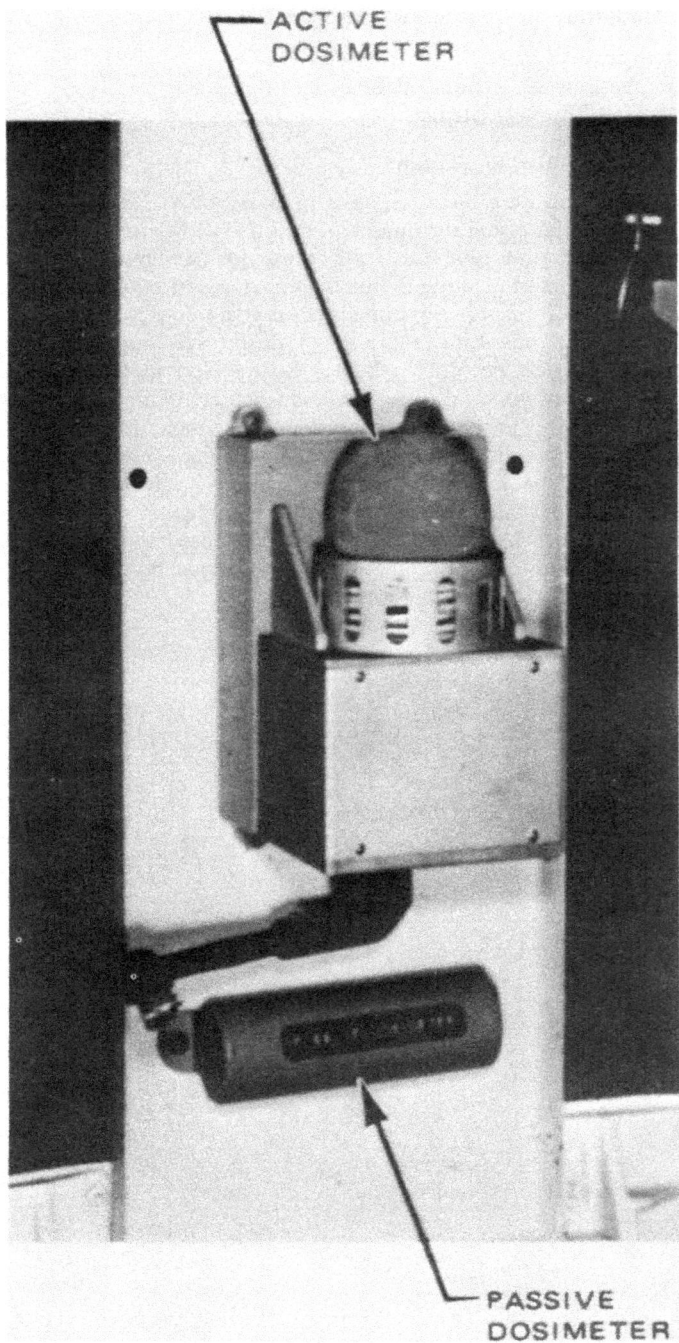

FIGURE 204.—Experiment D008, Instrumentation to measure radiations inside Skylab.

D024, Thermal Control Coatings (Fig. 205)

Principal Investigator:
  Dr. William Lehn
  Wright-Patterson Air Force Base
  Dayton, Ohio

Objective and Instrumentation:

Expose samples of thermal control coating materials to the space environment in order to compare results with ground-based simulations, and to determine mechanisms of degradation due to the space environment and space radiation. This experiment measures degradation that occurs only while the Skylab is in or near Earth orbit; Experiment M415 measures

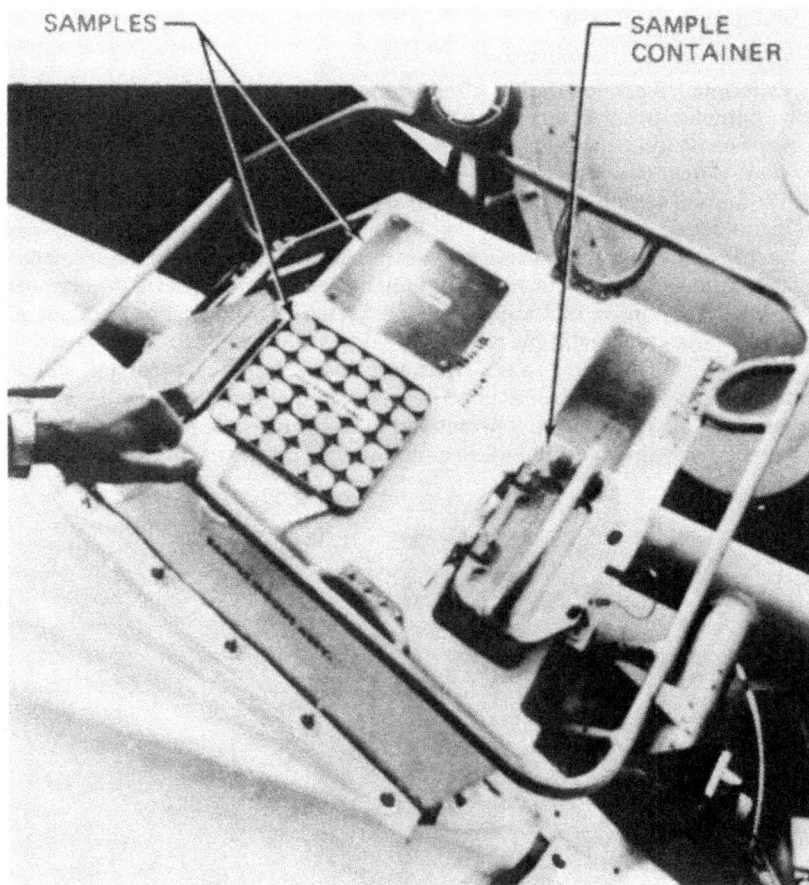

FIGURE 205.—Experiment D024, Degradation of thermal control coatings in a near-Earth space environment.

launch and pre-launch effects. The instrumentation consists of two panels, each containing 36 thermal control coating samples (Discs of 2.5 cm or 1 inch diameter). The panels will be protected by covers and will be exposed only to the space environment. One panel will be retrieved and placed in a hermetically sealed container and then returned to Earth for analysis of the first manned mission; the other will be returned to Earth on the second manned mission.

M415, Thermal Control Coatings (Fig. 206)

Principal Investigator:
  Eugene C. MacKannan
  NASA-Marshall Space Flight Center
  Huntsville, Alabama

Objective and Instrumentation:

Determine the degradation effects of prelaunch, launch, and space environments on the thermal absorption and emission characteristics of various coatings commonly used for passive thermal control. The principal elements of this experiment consist of two panels, each containing 12 thermal sensors arranged in four sets of three, mounted on the exterior of the Saturn IB launch vehicle. Three different thermal control coating samples are mounted on the sensors in each set. One set of sample coatings will be exposed to all environments. A second set will be exposed immediately prior to Launch Escape System Tower jettison and to all environmental conditions thereafter. The third set will be exposed to retrorocket firing and space environments, while the fourth set will be exposed to space environment only. Average thermal-radiative properties can be calculated from telemetered temperature measurements. These calculated values will then indicate how the various environments altered

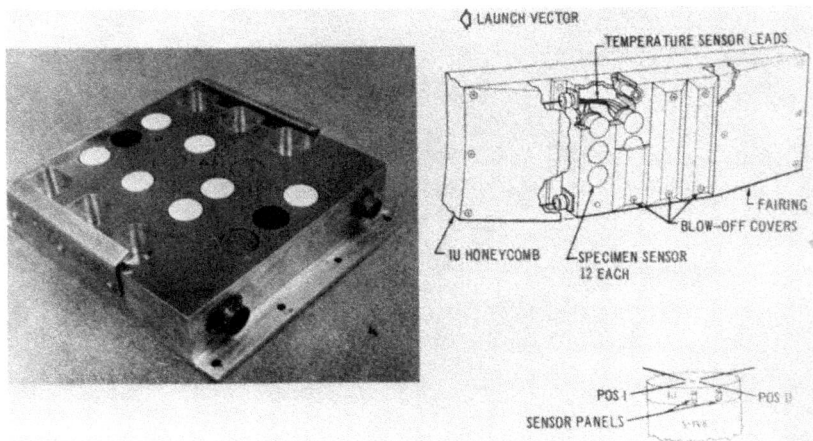

FIGURE 206.—Experiment M415, Degradation of thermal control coatings, before, during, and after launch.

coating characteristics. Unlike Experiment D024, detailed spectral reflection measurements cannot be made in this experiment since the coatings will not be retrieved.

T003, Inflight Aerosol Analysis (Fig. 207)

Principal Investigator:
Dr. William Z. Leavitt
Department of Transportation
Cambridge, Massachusetts

Objectives and Instrumentation:

Measure the size, concentration, and composition of the minute particles present in the atmosphere inside Skylab. The information obtained will be used not only as a measure of the atmospheric quality in Skylab but as a data source in analysis of other Skylab phenomena. Sources of astronaut discomfort, either respiratory or skin, may be related to aerosol buildup; system performance anomalies may be resolved using these aerosol experiment data; also, the data can be used in the design of future spacecraft and equipment. The experiment is self-contained in a box approximately 15 cm × 25 cm × 33 cm (6 in. × 10 in. × 13 in.) with an air inlet, an air outlet, a filter selector knob, a channel indicator, and a particle-count readout register (Fig. 207). The channel indicator reads "1", "2", or "3", and the register gives the corresponding concentration of particles in the 1.0 to 3.0 micrometer, 3.0 to 9.0 micrometer, and 9.0 to 100 micrometer ranges. The different settings of the filter are for the different locations where measurements are made. The filter is used to bring particles back for later identification and position correlation. Particle size and count are determined by passing a known volume of

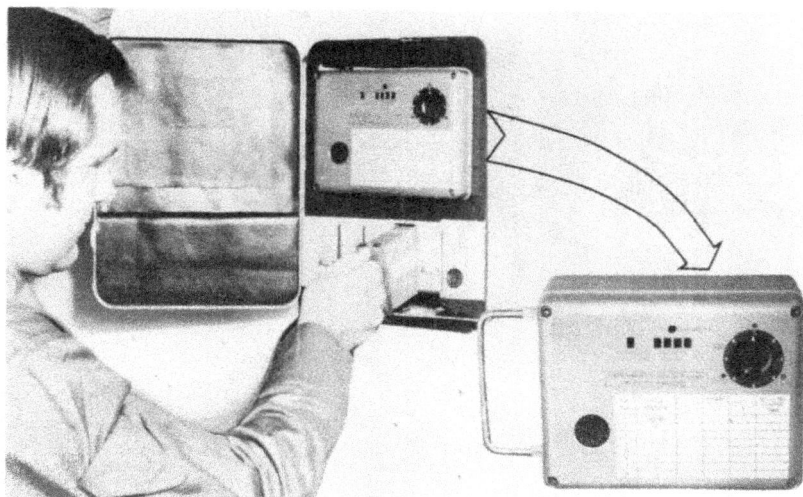

FIGURE 207.—Experiment T003 to determine the presence and number of dust and other particles (aerosols) in the cabin atmosphere.

air through the measuring chamber and measuring the amount of light each particle scatters to a photodetector, and also the number of light pulses which corresponds to the number of particles. Measurements at a main location will be taken three times a day. Every ten days, measurements are made at several other predetermined stations. Ten more measurements can be made anywhere at any time at the crew's discretion. The filters used in each measurement will be returned to Earth for analysis.

T025, Coronagraph Contamination Measurements (Fig. 208)

Principal Investigator:
  Dr. Mayo Greenberg
  Dudley Observatory
  Albany, New York

Objectives and Instrumentation:

Observe visually and photographically and interpret the particulate atmosphere surrounding the Skylab in orbit; study the change (size, quantity, distribution) in these particles caused by thruster firing and water dumps; and photograph the solar corona to find out if the contamination degrades the ability to see. A disc assembly, mounted on a boom, is extended through the Solar Scientific Airlock. A camera is mounted on the T025 canister so it remains inside the OWS. The disc assembly occults the solar disc from the camera lens so the corona may be photo-

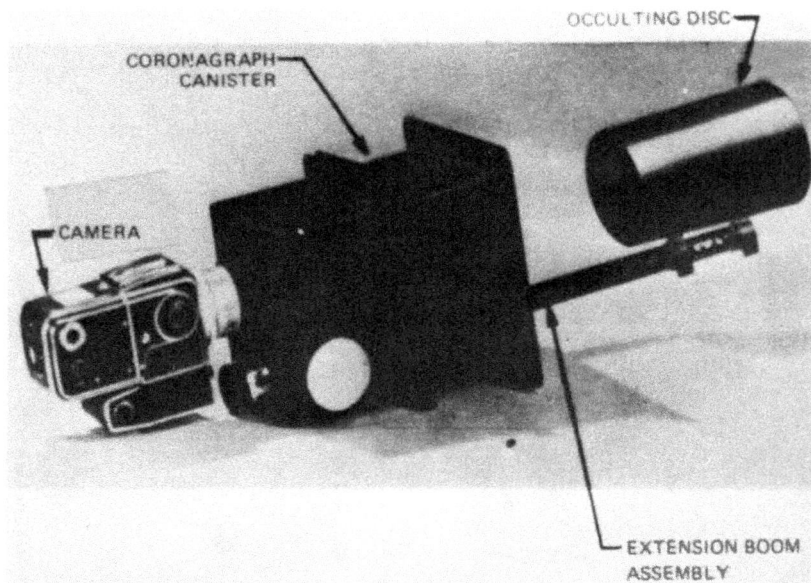

FIGURE 208.—Experiment T025, Measurement of contamination near Skylab with a coronagraph.

graphed. Of essential interest is the amount of background "fog" produced by stray light falling on the film during exposure. Film will be returned to Earth for analysis.

T027, ATM Contamination Measurements (Fig. 209)

Principal Investigator:
Dr. Joseph A. Muscari
Martin-Marietta Corporation
Denver, Colorado

Objective and Instrumentation:

Determine the change in optical properties of various surfaces caused by contaminants near the spacecraft by postflight analysis, measure the amount of contaminants deposited on a test surface during flight, and observe the sky brightness caused by solar illumination of contaminants. A sample array system containing 200 samples of 16 different materials (windows, mirrors, gratings, other optical surfaces) and two quartz crystal microbalances will be exposed to the space environment for various durations during a period of five days. The quartz crystal microbalances will measure contaminant masses deposited on their surfaces in near real time. A photoelectric polarimeter photometer, used jointly for Experiments S073 and T027, will measure the brightness of contaminants illuminated by the Sun at elongations longer than 15° from the Sun. A 16 mm camera will be used to record the photometric and polarimetric data on film. At the end of the mission, the individual samples on the sample array will be covered, and the sample array will be placed in a vacuum container for return to Earth for postflight analysis.

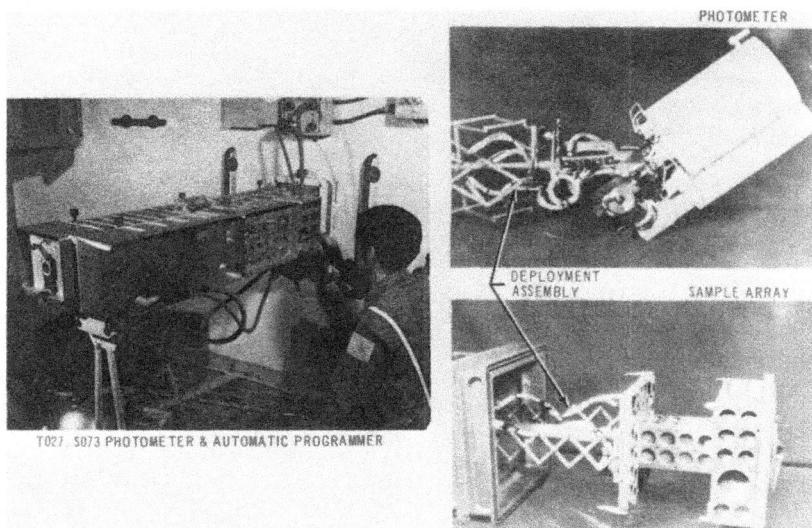

FIGURE 209.—Experiment T027, Measurement of contamination near Skylab with surface samples.

5. SKYLAB STUDENT PROJECT

Through the Skylab Student Project, high school students of the United States were given the opportunity to participate in the Skylab scientific program. All students in the ninth through the twelfth grades in all United States public, private, parochial, and overseas schools were eligible. The Project's purpose is to stimulate interest in science and technology by directly involving secondary school students in a space research program.

In October 1971, the National Science Teachers Association, under the auspices of NASA, distributed announcements of the science opportunity and of the method of participating in the Skylab Program. As a result, over 3,400 proposals for experiments were received. The National Science Teachers Association then selected 25 proposals as the national winners, announcing their names in April, 1972. The 25 national finalists from 16 states took part in a week of preliminary design review at the George C. Marshall Space Flight Center where they, their parents, and their teacher/sponsors were joined by Skylab scientists, engineers, technicians, and project officials.

After a detailed review by NASA, 19 out of the 25 experiments selected as national winners were approved for Skylab. The NASA review determined that, because of Skylab performance requirements and schedule restrictions, the other six experiments could not be accommodated. This experiment evaluation and flight selection process involved NASA Skylab personnel from the Marshall Space Flight Center, the Johnson Space Center, and the Kennedy Space Center. The 19 students whose experiments were selected for research participated closely in the development of necessary experiment instrumentation and in the detailed planning of their investigations including data retrieval and processing, flight planning, and crew training. After the mission, the students will evaluate their data and report on their experiments. These student experiments, which are handled by Skylab project management in a manner very similar to the handling of main Skylab experiments, reflect remarkable technical abilities. The names of the student experimenters, description of their experiments, and the experiment numbers follow:

JOE B. ZMOLEK, Oshkosh, Wisconsin
Lourdes High School, Mr. William L. Behring, Teacher/Sponsor
"Absorption of Radiant Heat in the Earth's Atmosphere," ED11.

This experiment is to derive information on the loss of heat energy in the Earth's atmosphere. It will utilize data from the Earth Resources Experiment Package, Experiment S191, and related ground-truth measurements to be made simultaneously at the Earth's surface.

TROY A. CRITES, Kent, Washington
Kent Junior High, Mr. Richard C. Putnam, Teacher/Sponsor
"Space Observation and Prediction of Volcanic Eruptions," ED12.

The purpose of this study is to analyze infrared surveys taken in the areas of known volcanoes by sensors of the Skylab Earth Resources Experiment

Package (Experiments S190A, S190B, S191, S192). The data will be compared with ground-based data to determine whether remote sensing can detect increased thermal radiation which may precede an imminent volcanic eruption.

ALISON HOPFIELD, Princeton, New Jersey
Princeton Day School, Mr. Norman Sperling, Director, Duncan Planetarium, Sponsor
"Photography of Libration Clouds," ED21.

The Skylab solar telescope cameras of the coronagraph experiment S052 will provide pictures of two regions in the Moon's orbit. At two points in the orbit of the Moon, ahead of and following the Moon in its path, a condition of gravitational equilibrium exists which causes the accumulation of space particles. When each of these regions comes within sight of the Skylab solar telescopes, pictures will be taken, and the brightness and polarization of the reflected light will be measured by Experiments T027 and S073.

DANIEL C. BOCHSLER, Silverton, Oregon
Silverton Union High School, Mr. John P. Dailey, Teacher/Sponsor
"Possible Confirmation of Objects within Mercury's Orbit," ED22.

This observation will attempt to identify a planetary body which may orbit the Sun at a distance approximately 0.1 of the distance from the Sun to the Earth (Mercury's orbital radius is about one-third of the radius of Earth's orbit). The experiment is to be performed by examining about 30,000 Skylab solar telescope photographs taken by the coronagraph of ATM Experiment S052.

JOHN C. HAMILTON, Aiea, Hawaii
Aiea High School, Mr. James A. Fuchigami, Teacher/Sponsor
"Spectrography of Selected Quasars," ED23.

Selected photographs obtained by the ultraviolet stellar astronomy equipment (S019) will be analyzed. Photographs of target areas in which quasars have been identified will be studied to obtain spectral data in the ultraviolet region to add to existing data in the radio and visible ranges.

JOE W. REIHS, Baton Rouge, Louisiana
Tara High School, Mrs. Helen W. Boyd, Teacher/Sponsor
"X-Ray Content in Association with Stellar Spectral Classes," ED24.

The primary aim is to make observations of celestial regions in X-ray wavelengths in an attempt to relate X-ray emissions to stars and to their spectral characteristics. In addition, observations of the Sun in X-ray and other spectral regions will be studied to re-examine the Sun and its relation to stars in other stellar classes. Pictures from ATM Experiments S054 and S056 will be evaluated.

JEANNE L. LEVENTHAL, Berkeley, California
Berkeley High School, Mr. Harry E. Choulett, Teacher/Sponsor
"X-Ray Emission from the Planet Jupiter," ED25.

The purpose of this study is to detect X-rays emitted from Jupiter. The X-ray emission, if detected by Skylab, will be correlated with solar activity and Jupiter's radio emission to derive more information on the radiation-emitting mechanisms of this planet. Pictures taken by the ATM Experiment S054 will be used.

NEAL W. SHANNON, Atlanta, Georgia
Fernbank Science Center, Dr. Paul H. Knappenberger, Teacher/ Sponsor
"A Search for Pulsars in Ultraviolet Wavelengths," ED26.

Ultraviolet observations of selected celestial regions will be used in an attempt to relate ultraviolet emissions with known radio-emitting pulsars, and with the pulsar in the Crab Nebula which is known to emit pulses in X-ray, visible light, and radio frequencies. Pictures taken by instrument S019 in the Stellar Astronomy Program will be evaluated.

ROBERT L. STAEHLE, Rochester, New York
Harley School, Mr. Alan H. Soanes, Teacher/Sponsor
"Behavior of Bacteria and Bacterial Spores in the Skylab Space Environment," ED31.

In this experiment, colonies of various species of bacteria will be studied in the Skylab zero-gravity environment to determine if this environment induces variations in survival rate, growth, and mutations of bacteria and spores different from those observed in identical colonies on Earth.

TODD A. MEISTER, Jackson Heights, New York
Bronx High School of Science, Mr. Vincent G. Galasso, Teacher/Sponsor
"An In-Vitro Study of Selected Isolated Immune Phenomena," ED32.

This experiment seeks to determine if the actions of antibodies are influenced by absence of gravity. An antibody is a substance which acts to destroy specific foreign substances (antigens) such as toxins, bacteria, and dust. Microscope slides used in the experiments are "fixed" by adding acetic acid to stop antibody activity resulting from the introduction of antigens. These slides are then photographed by the astronaut; the photographs are returned to Earth for analysis by the students.

KATHY L. JACKSON, Houston, Texas
Clear Creek High School, Mrs. Mary K. Kimzey, Teacher/Sponsor
"A Quantitative Measure of Motor Sensory Performance During Prolonged Flight in Zero Gravity," ED41 (Fig. 210).

This experiment uses a standard eye-hand coordination test apparatus to measure changes in motor sensory skill of crew members. An astronaut has to make electric contacts with a hand-held stylus through a pattern of holes in a punch-board-like plate. Total time need for this procedure is a measure of coordination efficiency. Results will be recorded on tape.

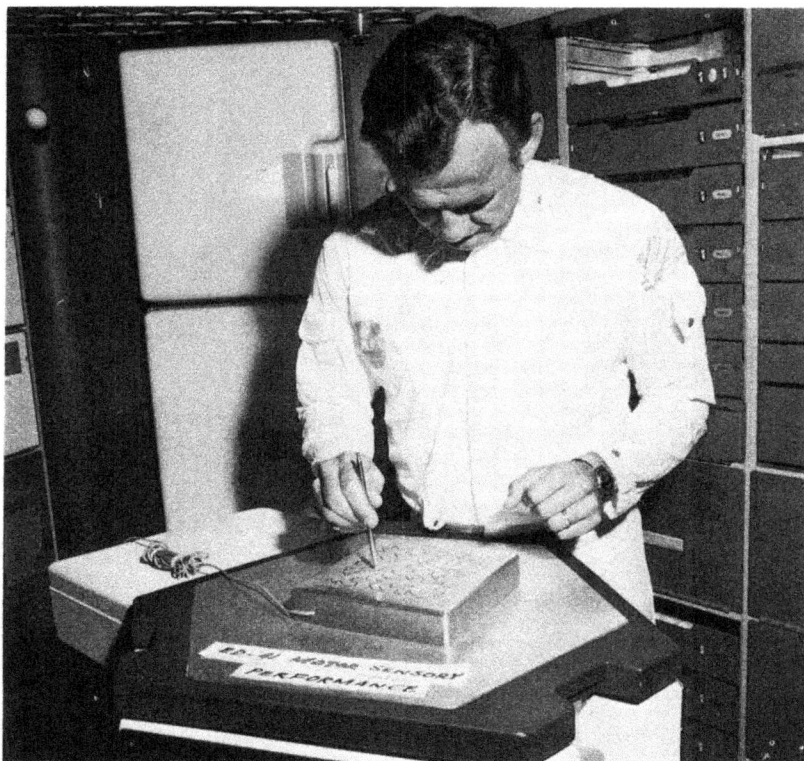

FIGURE 210.—Experiment ED41, Study of motor sensory performance with punch board and stylus.

JUDITH S. MILES, Lexington, Massachusetts
Lexington High School, Mr. J. Michael Conley, Teacher/Sponsor
"Web Formation in Zero Gravity," ED52 (Fig. 211).

Photographic observations will be made of the web building process and the detailed structure of the web of the common cross spider (*araneus diadematus*) in an Earth environment and in the weightless Skylab environment. Analysis of experiment results will be similar to that of like experiments, without the Skylab environment, performed by the Research Division, North Carolina Department of Mental Health.

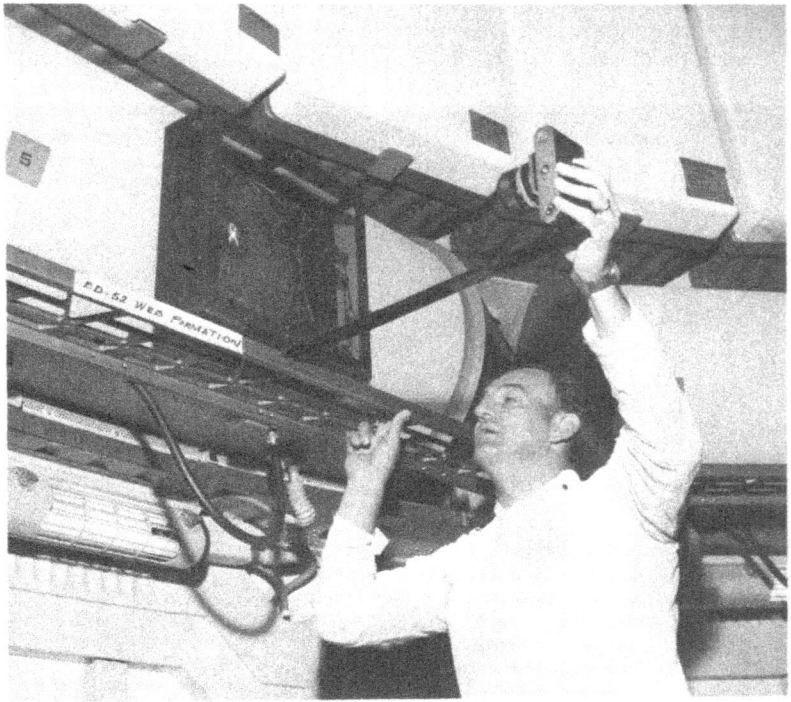

JOEL G. WORDEKEMPER, West Point, Nebraska
Central Catholic High School, Mrs. Louis M. Schaaf, Teacher/Sponsor
"Plant Growth in Zero Gravity," ED61.
DONALD W. SCHLACK, Downey, California
Downey High School, Miss Jean C. Beaton, Teacher/Sponsor
"Photographic Orientation of an Embryo Plant in Zero Gravity," ED62.

These two experiments have been combined into a single joint experiment whose objectives are:

1. To determine the differences between rice seedlings grown in zero-gravity, and rice seedlings grown on Earth under similar conditions, with respect to root and stem growth, and orientation.

2. To determine whether light can be used as a substitute for gravity in causing the roots and stems of rice seedlings to grow in a desired direction under zero-gravity, and if so, to determine the minimum light level required.

Photographic records of the experiment will be returned to Earth.

CHERYL A. PELITZ, Littleton, Colorado
Arapahoe High School, Mr. Gordon B. Scheels, Teacher/Sponsor
"Cytoplasmic Streaming in Zero-Gravity," ED63 (Fig. 212).

Microscopic observation will be performed by an astronaut on leaf cells
of elodea plants [1] in zero gravity to determine if there is a difference between
the motion of intracellular cytoplasm under weightlessness and intercellular
cytoplasmic motion in similar leaf cells on Earth. Cytoplasm is the protoplasm
of a cell exclusive of the nucleus.

---

[1] Eloda is a bright green, fast growing plant found in fresh water ponds.

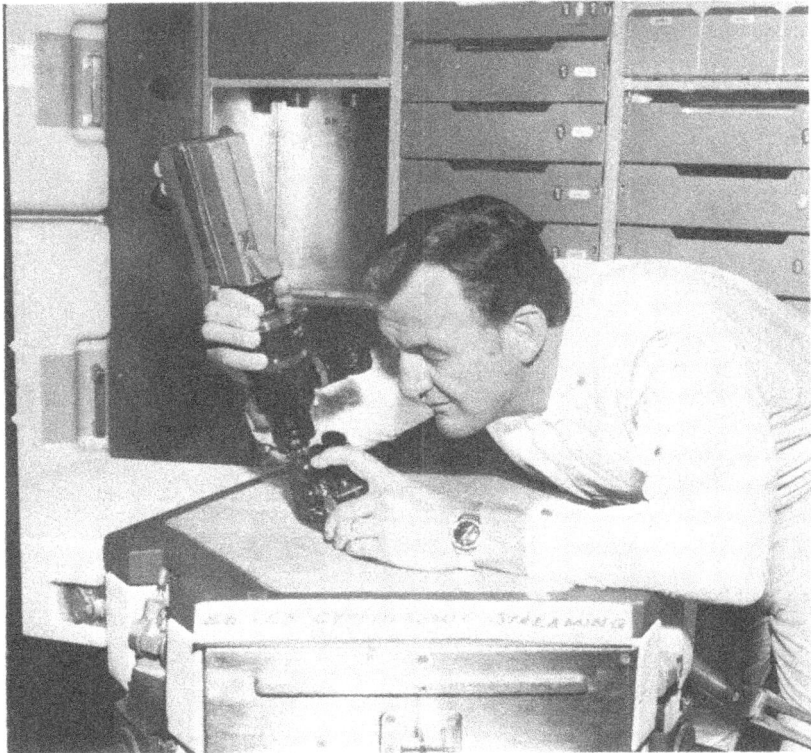

FIGURE.212.—Experiment ED63, Observation of cytoplasmic streaming in zero gravity.

500-721 O - 73 - 15

ROGER G. JOHNSTON, St. Paul, Minnesota
Alexander Ramsey High School, Mr. Theodore E. Molitor, Teacher/
    Sponsor
"Capillary Action Studies in a State of Free Fall," ED72 (Fig. 213).

The purpose of this experiment is to determine if the zero-gravity environment induces changes in the characteristics of capillary and wicking action from the familar Earth-gravity characteristics. The motion of liquids through capillary tubes will be recorded photographically.

FIGURE 213.—Experiment ED72, Study of capillary action under zero gravity.

VINCENT W. CONVERSE, Rockford, Illinois
Harlem High School, Miss Mary J. Trumbauer, Teacher/Sponsor
"Zero-Gravity Mass Measurement," ED74 (Fig. 214).

This experiment complements the existing Skylab specimen mass and body mass measurement devices. The equipment consists of a simple leaf spring anchored at one end and with the mass to be measured at the other end. The experiment operates on the same principle as the biomedical Skylab mass measurement devices.

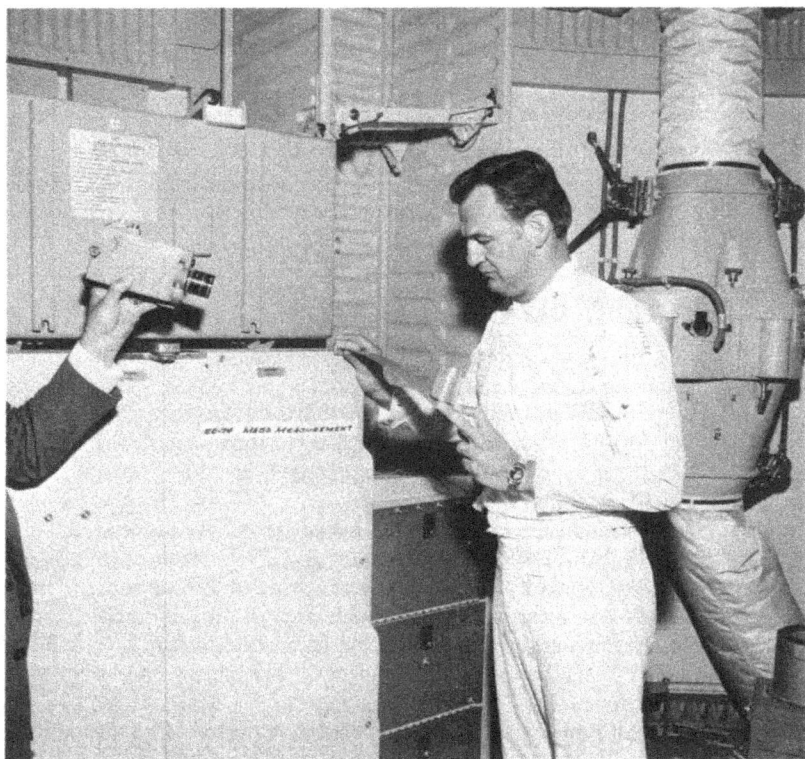

FIGURE 214.—Experiment ED74, Mass measurement under zero gravity with a leaf spring system.

TERRY C. QUIST, San Antonio, Texas
Thomas Jefferson High School, Mr. Michael Stewart, Teacher/Sponso
"Earth Orbital Neutron Analysis," ED76

Detectors inside Skylab will record neutrons from three potential source
Earth albedo neutrons, high energy neutrons from the Sun, and neutrons fr
secondary processes on Skylab. The detectors mounted on the inboard fac
of the Skylab water tanks will record neutrons which have been moderat
during their passage through the water in the tanks, and then produce fissi
particles which generate ionization tracks in a plastic material. Detect
mounted elsewhere in the Skylab will furnish control data. Chemical tre
ment of the plastic after return to Earth will reveal readily identifiable trac

W. BRIAN DUNLAP, Youngstown, Ohio
Austintown Fitch High School, Mr. Paul J. Pallante, Teacher/Sponsor
"Wave Motion through a Liquid in Zero-Gravity," ED78.

This experiment will observe the motion of a gas bubble surrounded b
fluid when excited by a known driving force. Provisions will be made for va
ing the size of the bubble.

6. POSTFLIGHT DATA EVALUATION

Responsibility for the reduction and evaluation of Skylab experiment d
lies primarily with the Principal Investigators. Telemetry data will be ma
available to investigators through the Mission Control Center at JSC
quickly as possible after receipt of the data. Films, tapes, samples, logboc
and other data records to be brought back in the returning Comma
Modules will be taken to the Mission Control Center by the recovery te
and then distributed to Principal Investigators. Additional data, such
spacecraft positional location, details of Skylab performance, trajectory
formation, Skylab attitude as a function of time and location, environmer
data, records of radiation and contamination sensors, and other data
specific interest to investigators will be furnished by the Lydon B. Johns
Space Center in Houston or the George C. Marshall Space Flight Cente
Huntsville.

Most of the experiments will be performed on all three manned missic
A preliminary "quick look" evaluation of experimental results after
first and the second mission will be of great importance in these experime
because results of one mission may determine whether an experiment sho
be modified for the following missions.

For some groups of experiments, one of the NASA Centers bears a br
overall responsibility; for example, the Lyndon B. Johnson Space Ce
is responsible for the Earth Resources Experiment Package (EREP), w
a number of Principal Investigators are individually responsible for
experiments within this package. Data evaluation plans for EREP incl
the following activities:

1. *By JSC (within two weeks after CM Recovery)* :
   Films will be developed.
   Films and tapes will be duplicated.

Strip charts, tabulations, and pictures will be made from tapes.
Logs and voice recordings will be transcribed.
Supporting data on Skylab performance will be documented.
Films and other records will be furnished to Principal Investigators.
Teams will make "quick look" evaluation.
Preliminary results will be furnished to Science Analysis Teams, Mission Planning, and NASA Public Affairs Offices.

2. *By the Principal Investigators:*

   a. Within four weeks after CM recovery:
      Detailed screening of data.
      Careful correlation of data with sensor performance and Skylab operational data.
      Data analysis and evaluation.
      Feedback made to NASA Mission Planning.
      Preliminary Science Reports written.

   b. During the first year after CM recovery:
      Thorough analysis of data.
      Data packages prepared for user agencies.
      Final Science Reports prepared.
      Suggestions made for future projects.

   c. After about one year:

Original data will be made available to any qualified investigator. Data evaluation plans for the various groups of Skylab experiments differ in some details; for example, ATM films will be developed by the Principal Investigators. However, the plan for EREP data described here reflects the characteristic features of all of the Skylab data evaluation plans.

CHAPTER VI

Ground-Based
Supporting Projects

Each of the experimental programs on Skylab represents an extension of studies that have been under way on Earth for some time. It is natural, therefore, that experimenters have expressed the desire to compare and correlate the results of Skylab experiments very carefully with data gained from ground-based observations. In fact, several ground-based study programs have been planned and prepared specifically to provide data in support of Skylab experiments. Ground-based supporting projects which will supply back-up services to Skylab will include the Ground-Based Astronomy Program, the National Oceanic and Atmospheric Administration's support to NASA, earth resources ground-truth activities, and ground-based medical studies.

These ground-based supporting projects will supplement Skylab studies in several ways. First, by comparing ground observations and Skylab observations of the same object, an experimenter can determine the kind and the degree of superiority which space observations may have over ground observations. Second, by observing a given object, such as a group of sunspots, over an extended period of time before and after the Skylab mission, insight into evolutionary processes will be gained which may greatly enhance the value of Skylab observations. Third, by observing at close range certain features on the surface of the Earth, such as tectonic formations, ocean currents, or plant growth patterns, a "calibration table" for the interpretation of Skylab pictures of the same features can be established. This calibration can then be used for the interpretation of other Skylab pictures for which no ground pictures exist. Fourth, in the case of biological and medical experiments, the ground-based observations will enable the investigator to isolate the effects of weightlessness upon living organisms by comparing Skylab data with the results of observations taken under similar conditions on the ground.

Details of some of the ground-based supporting projects will be described in the following sections.

1. GROUND-BASED ASTRONOMY PROGRAM

The Skylab Ground-Based Astronomy Program is designed to obtain solar data from observatories around the world at the same time the ATM instruments are observing the Sun from orbit. Data gathered on the ground will support and add to the space-gathered data and thus help gain maximum benefit from the ATM.

Early in 1971, NASA began to solicit proposals from solar astronomers for ground-based observations that would support and extend solar observations on Skylab. Many organizations submitted proposals; nine of them were selected for implementation.

The participating astronomical observatories and their proposed projects are as follows:

a. The University of Hawaii's Institute of Astronomy at Haleakala (Fig. 215) is constructing a photometer for observations of active regions in the corona. This instrument will measure simultaneously the intensities of several visible coronal lines in a study to determine the rates of energy loss and gain from the active regions and the effects of flare events on the corona.

b. The Lockheed Missiles and Space Company at Palo Alto, California, will operate at Kitt Peak's McMath solar observatory (Fig. 216) a spectroheliograph capable of mapping physical parameters of the solar atmosphere. In conjunction with the Kitt Peak solar telescope and vacuum spectograph, a wide-exit aperture and a specially constructed movie camera with rapid film advance will be used to obtain spectral maps of active regions of the solar disc with high spatial resolution (to one-half arc second). A high speed microdensitometer-computer system will be set up to allow rapid analysis of the spectral data to obtain solar plasma velocity and magnetic field maps of regions of interest.

c. The National Bureau of Standards is upgrading calibration capabilities in support of ATM-related measurements in the following areas:

(1) A hydrogen arc source of known radiant flux for the calibration of spectrometric detector systems over the region from 50 to 370 nm (500 to 3700 Å) is being developed.

(2) A study is being conducted to determine the effects on photocathodes caused by removal or addition of mono-layers of contaminants in vacuum; wavelength region of interest is 50 to 150 nm (500 to 1500 Å).

(3) A capability is being established for radiometric calibration down to 20 nm (200 Å) by utilizing the National Bureau of Standards' synchrotron facility. Windowless diodes are being developed as transfer standards in this spectral range.

FIGURE 215.—Solar Observatory on Mount Haleakala, Hawaii.

FIGURE 216.—Solar Observatory on Kitt Peak, Arizona. Courtesy of Kitt Peak National Observatory.

d. California Institute of Technology is installing a 65 cm (25.6 in) photoheliograph at its observatory at Big Bear Lake in California (Fig. 217) where observing conditions are exceptionally good. The new unit will be used for extremely high resolution studies of active regions in conjunction with ATM. Filtergrams will be made in lines extending from the 393.3 nm (3933 Å) Calcium K-line to the 1083 nm. (10,830 Å) Helium line.

e. Lockheed Solar Observatory at Rye Canyon, California (Fig. 218) is preparing both of its spar telescope systems for observations during the ATM mission. On one spar, studies in the D_3 line of Helium I will be directed at solar flares and transient events during periods of high disc activity and to prominence observations during periods of low disc activity. On the other spar, a telescope with a filter will make high resolution photographic studies in the calcium II line at 854.2 nm (8542 Å). This line is believed to originate at a higher level in the solar atmosphere than the H-Alpha line; it should be very valuable in relating ATM X-ray and extreme UV data to filtergrams and spectroheliograms taken at wave lengths originating at lower levels in the chromosphere.

f. The University of California at San Diego is building a photometer system for infrared observations of solar flares at the Mt. Lemmon Observatory near Tucson, Arizona (Fig. 219). Observations with the 1.52 m (60-in) Cassegrainian telescope will extend from the 700 micrometer (700,000 nm) region down to possibly one micrometer. These infrared continuum observations will be compared with observations made on ATM.

FIGURE 217.—Solar Observatory at Big Bear Lake, California.

FIGURE 218.—Lockheed Solar Observatory at Rye Canyon, California.

g. The Lockheed Missile and Space Company at Palo Alto, California, will perform a theoretical study of Helium emissions in the visible and ultraviolet from solar active regions. Calculations of the statistical equilibrium populations for a 19 level Helium I and a 10 level Helium II ion are being performed. Results of these studies should allow the interpretation of line observations by ATM in terms of electron density and temperature profiles for the active regions on the Sun.

h. The Uttar Pradesh State Observatory in Naini Tal, India, is performing a study of dissociation and excitation equilibria of various molecules in the photosphere, in Sun spots, and in faculae. Detection equipment has been loaned to India for an observational program with its existing horizontal solar telescope and associated spectrograph. The program aims at an improvement of existing models of the solar atmosphere.

i. The Applied Physics Laboratory of Johns Hopkins University in Baltimore, Maryland (Fig. 220), will perform an observational program during the ATM mission of solar radio bursts in the 500 to 1000 MHz range. With the .01 second time resolution of the spectrograph and the 18.3 m (60-ft.) diameter antenna, it should be possible to determine whether a finite frequency drift rate is present in the radio bursts. This in turn will have implications on existing models of emission mechanism and electron densities in the source region. Results of these studies will be correlated with ATM data.

2. NATIONAL OCEANIC AND ATMOSPHERIC ADMINISTRATION SUPPORT TO ATM

The National Oceanic and Atmospheric Administration (NOAA) of the Federal Government has been coordinating a solar data collection network among observatories in the U.S. and in foreign countries. These observatories will furnish data to NOAA for use in its continuing study of solar activities.

FIGURE 219.—Mt. Lemmon Observatory, Tucson, Arizona.

FIGURE 220.—Radio Telescope of Johns Hopkins University, Baltimore, Maryland.

By agreement with NASA, the Space Environment Laboratory of NOAA established the Space Environment Support Program. This program is designed to support the unique requirements of the ATM Skylab mission for solar data. It will expand and improve the Space Environment Laboratory's current data gathering and distribution facilities in order to provide continuous solar activity forecasts to the ATM Principal Investigators at the Lyndon B. Johnson Space Center during the Skylab missions.

NOAA will use information from numerous observatories to provide this support to NASA before, during, and after Skylab flights. On the basis of solar data from the ground-based network of observatories, NOAA will forecast flares and other solar activities, inform NASA of important solar events in progress, and, after the Skylab mission, prepare a book of all data collected during the mission for use in data analysis.

During the mission, NOAA representatives will be stationed at the Lyndon B. Johnson Space Center to provide NASA with real-time space environment data, analyses, and forecasts. These representatives will coordinate the operation of the solar data network. Working closely with the Principal Investigators and providing them with current information on the present state of the Sun, they will interpret data for the Principal Investigators and make suggestions on possible observing programs based on solar activity. After the mission, NOAA will continue collection of data from the network for a month.

225

3. EARTH RESOURCES GROUND TRUTH ACTIVITIES

"Ground truth" data will be obtained by direct observations on the ground of those areas, objects, and phenomena which will also be observed by Skylab instruments from orbit. Nearly simultaneous observations will be made of weather, lighting conditions, and other environmental factors which may influence the data gathered from Skylab. By comparing ground truth observations with orbital observations of a test site, calibration factors will be established which will allow the proper interpretation of orbital data from many sites.

The concept to be followed in obtaining ground truth data to correlate with Skylab observational data from the Earth Resources Experiment Package is to employ each individual Principal Investigator's own ground truth capability. Principal Investigators will furnish their ground truth data to the Skylab Ground Truth Office at JSC where they will be included in the archives together with the space-acquired data.

The Lyndon B. Johnson Space Center will establish a communications system so that Principal Investigators will be notified of the times Skylab will overfly their areas. Thus, they can acquire ground truth measurements synchronized with EREP overflight data.

Ground truth data to be acquired in support of EREP observations will include photographs, temperature measurements, terrain data, weather observations, and descriptive material identifying plant growth, soil conditions, snow depth, status of crops, geological features, and other specific details.

4. AIRCRAFT UNDERFLIGHT SUPPORT

NASA-operated and private aircraft will be used to obtain data over the sites to be observed by Skylab. These aircraft will carry a variety of cameras and imaging devices which generally approximate the capabilities available on board Skylab. Like the ground truth data, data acquired by high altitude aircraft underflights will be used to analyze and understand the space-acquired data.

CHAPTER VII

Summary and Outlook

Skylab marks a transition in the American space flight program. It represents a step from small to large satellites, from short manned flights to longtime manned orbiters, from the astronauts' role as observers in space to the role of workers in space, from single-purpose spacecraft to multipurpose space stations, from a time of space exploration to a time of space utilization. It stands at the beginning of an era in which near-Earth space with its weightlessness and its superior viewing conditions is offering a work place for man that would never be available under the restricted conditions on the ground. It will provide the basis for technological developments and for scientific research which may become decisive in man's struggle with the growing problems of his evolving civilization.

The operational life of Skylab will extend over a period of eight months. Analysis and evaluation of the numerous measurements taken on Skylab will cover a far longer period. Results of Skylab experiments, besides enriching our knowledge in technology and science, will undoubtedly influence many other space projects to be developed by the United States and abroad during forthcoming years. A new generation of manned orbiting flights, again expanding our capabilities in space and the utility of space flight for man on Earth, will begin with the Space Shuttle Program around the end of the 1970's. The Shuttle Program, too, will profit decisively from Skylab, our first station in space.

CHAPTER VIII

Listing of Skylab Experiments

Number	Title	Location on Skylab	Principal investigator	Manned mission			Responsibility
				1	2	3	
	Solar Studies						
S020	Ultraviolet and X-Ray Solar Photography	OWS	Dr. R. Tousey, U.S. Naval Research Laboratory		X	X	JSC
S052	White-Light Coronagraph	ATM	Dr. R. MacQueen, High Altitude Observatory	X	X	X	MSFC
S054	X-Ray Spectrographic Telescope	ATM	Dr. R. Giacconi, American Science and Engineering Corp. Acting: Dr. G. Vaiana.	X	X	X	MSFC
S055	UV Scanning Polychromator Spectroheliometer	ATM	Dr. L. Goldberg, Kitt Peak National Observatory. Acting: Dr. E. Reeves, Harvard College Observatory.	X	X	X	MSFC
S056	X-Ray Telescope	ATM	J. E. Milligan, MSFC	X	X	X	MSFC
S082A	Extreme UV Spectroheliograph	ATM	Dr. R. Tousey, U.S. Naval Research Laboratory	X	X	X	MSFC
S082B	Ultraviolet Spectrograph	ATM	Dr. R. Tousey, U.S. Naval Research Laboratory	X	X	X	MSFC
	Stellar Astronomy						
S019	UV Stellar Astronomy	OWS	Dr. K. G. Henize, JSC	X	X		JSC
S150	Galactic X-Ray Mapping	IU	Dr. W. L. Kraushaar, University of Wisconsin			X	MSFC
S183	UV Panorama Telescope	OWS	Dr. G. Courtès, Laboratoire d'Astronomie Spatiale, France	X	X		MSFC

See footnotes at end of table.

Space Physics

No.	Experiment	Module	Principal Investigator				Center
S009	Nuclear Emulsion Package	MDA	Dr. M. M. Shapiro, U.S. Naval Research Laboratory	X	----	----	MSFC
S063	UV Airglow Horizon Photography	OWS	Dr. D. M. Packer, U.S. Naval Research Laboratory	X	----	----	JSC
S073	Gegenschein and Zodiacal Light	OWS	Dr. J. L. Weinberg, Dudley Observatory	X	X	X	MSFC
S149	Micrometeoroid Particle Collection	OWS	Dr. C. L. Hemenway, Dudley Observatory	X	X	X	JSC
S228	Transuranic Cosmic Rays	OWS	Dr. P. B. Price, University of California at Berkeley	X	X	X	MSFC
S230	Magnetospheric Particle Composition	ATM	Dr. D. L. Lind, JSC, and Dr. Johannes Geiss, University of Bern, Switzerland.	X	X	X	MSFC

Earth Resources Experiments

No.	Experiment	Module	Principal Investigator				Center
S190A	Multispectral Photographic Cameras	MDA	K. Demel, Project Scientist, JSC	X	X	X	JSC
S190B	Earth Terrain Camera	OWS	K. Demel, Project Scientist, JSC	X	X	X	JSC
S191	Infrared Spectrometer	MDA	Dr. T. L. Barnett, JSC	X	X	X	JSC
S192	Multispectral Scanner	MDA	Dr. C. K. Korb, JSC	X	X	X	JSC
S193	Microwave Radiometer/Scatterometer and Altimeter.	MDA	D. E. Evans, JSC	X	X	X	JSC
S194	L-Band Radiometer	MDA	D. E. Evans, JSC	X	X	X	JSC

Life Sciences Projects

No.	Experiment	Module	Principal Investigator				Center
M071	Mineral Balance	OWS	G. D. Whedon, M.D., National Institutes of Health, and L. Lutwak, M.D., Cornell University.	X	X	X	JSC
M073	Bioassay of Body Fluids	OWS	Dr. C. S. Leach, JSC	X	X	X	JSC
M074	Specimen Mass Measurement	OWS	W. E. Thornton, M.D., JSC, and J. W. Ord, Col., Medical Corps, Clark AFB.	X	X	X	JSC
M078	Bone Mineral Measurement	None	J. M. Vogel, M.D., U.S. Public Health Service Hospital, San Francisco, and Dr. J. R. Cameron, University of Wisconsin Medical Center.	(Preflight and postflight)			JSC
M092	Lower Body Negative Pressure Device	OWS	R. L. Johnson, M.D., JSC, and J. W. Ord, Col., Medical Corps, Clark AFB.	X	X	X	JSC
M093	Vectorcardiogram	OWS	N. W. Allebach, M.D., USN Aerospace Medical Institute, and R. F. Smith, M.D., School of Medicine, Vanderbilt University.	X	X	X	JSC
M111	Cytogenic Studies of the Blood	None	L. H. Lockhart, M.D., University of Texas Medical Branch, Galveston, and P. C. Gooch, Brown and Root-Northrop.	(Preflight and postflight)			JSC

See footnotes at end of table.

Number	Title	Location on Skylab	Principal investigator	Manned mission 1	Manned mission 2	Manned mission 3	Responsibility
M112	Man's Immunity, In-Vitro Aspects	OWS	S. E. Ritzmann, M.D., and W. C. Levin, M.D., both of University of Texas Medical Branch, Galveston.	X	X	X	JSC
M113	Blood Volume and Red Cell Life Span	OWS	P. C. Johnson, M.D., Baylor University College of Medicine	X	X	X	JSC
M114	Red Blood Cell Metabolism	OWS	C. E. Mengel, M.D., University of Missouri, School of Medicine	X	X	X	JSC
M115	Special Hematological Effects	OWS	Dr. S. L. Kimsey and C. L. Fischer, M.D., JSC	X	X	X	JSC
M131	Human Vestibular Function	OWS	A. Graybiel, M.D., and Dr. E. F. Miller, both of USN Aerospace Medical Institute.	X	X	----	JSC
M133	Sleep Monitoring	OWS	J. D. Frost, Jr., M.D., Baylor University College of Medicine.	X	X	----	JSC
M151	Time and Motion Study	OWS	Dr. J. F. Kubis, Fordham University and Dr. E. J. McLaughlin, NASA Hq. OMSF.	X	X	X	JSC
M171	Metabolic Activity	OWS	E. L. Michel and Dr. J. A. Rummel, JSC	X	X	X	JSC
M172	Body Mass Measurement	OWS	W. E. Thornton, M.D., JSC	X	X	X	JSC
S015	Effect of Zero-Gravity on Single Human Cells	CM	P. O. Montgomery, M.D., and Dr. J. Paul, both of University of Texas Southwestern Medical School at Dallas.	X	----	----	JSC
S071	Circadian Rhythm, Pocket Mice	CSM	Dr. R. G. Lindberg, Northrop Corporation Laboratories		X	----	ARC
S072	Circadian Rhythm, Vinegar Gnats	CSM	Dr. C. S. Pittendrigh, Stanford University		X	X	ARC

Material Science & Manufacturing in Space

Number	Title	Location on Skylab	Principal investigator	Manned mission 1	Manned mission 2	Manned mission 3	Responsibility
M479	Zero-Gravity Flammability	MDA	J. H. Kinzey, JSC			X	MSFC
M512	Materials Processing Facility	MDA	P. G. Parks, MSFC			X	MSFC
(M551)	Metals Melting	MDA	R. M. Poorman, MSFC	X			MSFC
(M552)	Exothermic Brazing	MDA	J. Williams, MSFC	X			MSFC
(M553)	Sphere Forming	MDA	E. A. Hasemeyer, MSFC	X			MSFC
(M555)	Gallium Arsenide Crystal Growth	MDA	Dr. M. Rubenstein, Westinghouse Electric Corporation	X			MSFC
M518	Multipurpose Electric Furnace System	MDA	A. Boese, MSFC (Project Engineer)			X	MSFC
(M556)	Vapor Growth of II-VI Compounds	MDA	Dr. H. Wiedemeir, Rensselaer Polytechnic Institute			X	MSFC
(M557)	Immiscible Alloy Compositions	MDA	J. Reger, Thompson Ramo Wooldridge			X	MSFC
(M558)	Radioactive Tracer Diffusion	MDA	Dr. T. Ukanwa, MSFC			X	MSFC
(M559)	Microsegregation in Germanium	MDA	Dr. F. Padovani, Texas Instruments			X	MSFC

See footnotes at end of table.

Code	Location	Description	Investigator			Center
(M560)	MDA	Growth of Spherical Crystals	Dr. H. Walter, University of Alabama in Huntsville		X	MSFC
(M561)	MDA	Whisker-Reinforced Composites	Dr. T. Kawada, National Research Institute for Metals, Japan		X	MSFC
(M562)	MDA	Indium Antimonide Crystals	Dr. H. Gatos, Massachusetts Institute of Technology		X	MSFC
(M563)	MDA	Mixed III-V Crystal Growth	Dr. W. Wilcox, University of Southern California, Los Angeles		X	MSFC
(M564)	MDA	Halide Eutectics	Dr. A. Yue, University of California at Los Angeles		X	MSFC
(M565)	MDA	Silver Grids Melted in Space	Dr. A. Deruythere, Catholic University of Leuven, Belgium		X	MSFC
(M566)	MDA	Copper-Aluminum Eutectic	E. Hasemeyer, MSFC		X	MSFC

Zero-Gravity Systems Studies

Code	Location	Description	Investigator			Center
M487	OWS	Habitability/Crew Quarters	C. C. Johnson, JSC	X	X	MSFC
M509	OWS	Astronaut Maneuvering Equipment	Maj. C. E. Whitsett, Jr., USAF Space & Missile Systems Organization	X	X	JSC

Spacecraft Environment

Code	Location	Description	Investigator			Center
M516	OWS	Crew Activities and Maintenance Study	R. L. Bond, JSC	X	X	JSC
T002	OWS	Manual Navigation Sightings	R. J. Randle, ARC	X	X	ARC
T013	OWS	Crew/Vehicle Disturbance	B. A. Conway, LaRC	X	X	LaRC
T020	OWS	Foot-Controlled Maneuvering Unit	D. E. Hewes, LaRC		X	LaRC
D008	CM	Radiation in Spacecraft	Capt. A. D. Grimm, USAF, Kirtland Air Force Base	X		AF, JSC
D024	AM	Thermal Control Coatings	Dr. W. Lehn, Wright-Patterson Air Force Base	X		AF, JSC
M415	IU	Thermal Control Coatings	E. C. McKannan, MSFC	X		MSFC
T003	OWS	Inflight Aerosol Analysis	Dr. W. Z. Leavitt, Department of Transportation	X	X	MSFC
T025	OWS	Coronagraph Contamination Measurements	Dr. M. Greenberg, Dudley Observatory	X	X	JSC
T027	OWS	ATM Contamination Measurements	Dr. J. A. Muscari, Martin-Marietta Corporation	X	X	MSFC

Skylab Student Project

Secondary School Student Winners

Code	Location	Description	Investigator			Center
ED11	None	Absorption of Radiant Heat in the Earth's Atmosphere	J. B. Zmolek, Oshkosh, Wisconsin	X	X	MSFC
ED12	None	Space Observation and Prediction of Volcanic Eruptions	T. A. Crites, Kent, Washington		X	MSFC
ED21	None	Photography of Libration Clouds	A. Hopfield, Princeton, New Jersey		X	MSFC
ED22	None	Possible Confirmation of Objects within Mercury's Orbit	D. C. Bochsler, Silverton, Oregon		X	MSFC

See footnotes at end of table.

Number	Title	Location on Skylab	Principal investigator	Manned mission 1	2	3	Responsibility
ED23	Spectrography of Selected Quasars	None	J. C. Hamilton, Aiea, Hawaii	X			MSFC
ED24	X-Ray Content in Association with Stellar Spectral Classes	None	J. W. Reihs, Baton Rouge, Louisiana			X	MSFC
ED25	X-Ray Emission from the Planet Jupiter	None	J. L. Leventhal, Berkeley, California		X		MSFC
ED26	A Search for Pulsars in Ultaviolet Wavelengths	None	N. W. Shannon, Atlanta, Georgia		X		MSFC
ED31	Behavior of Bacteria and Bacterial Spores in the Skylab Space Environments	OWS	R. L. Staehle, Rochester, New York	X			MSFC
ED32	As In-Vitro Study of Selected Isolated Immune Phenomena	OWS	T. A. Meister, Jackson Heights, New York		X		MSFC
ED41	A Quantitative Measure of Motor Sensory Performance During Prolonged Flight in Zero Gravity	OWS	K. L. Jackson, Houston, Texas			X	MSFC
ED52	Web Formation in Zero Gravity	OWS	J. S. Miles, Lexington, Massachusetts		X		MSFC
ED61	Plant Growth in Zero Gravity	OWS	J. G. Wordekemper, West Point, Nebraska			X	MSFC
ED62	Phototropic Orientation of an Embryo Plant in Zero Gravity	OWS	D. W. Schlack, Downey, California			X	MSFC
ED63	Cytoplasmic Streaming in Zero Gravity	OWS	C. A. Peltz, Littleton, Colorado		X		MSFC
ED72	Capillary Action Studies in a State of Free Fall	OWS	R. G. Johnson, St. Paul, Minnesota			X	MSFC
ED74	Zero Gravity Mass Measurement	OWS	V. W. Converse, Rockford, Illinois		X		MSFC
ED76	Earth Orbital Neutron Analysis	OWS	T. C. Quist, San Antonio, Texas	X	X		MSFC
ED78	Wave Motion Through a Liquid in Zero Gravity	OWS	W. B. Dunlap, Youngstown, Ohio		X	X	MSFC

AF—Air Force.
A RC—Ames Research Center, Moffett Field, Calif.
LaRC—Langley Research Center, Hampton, Va.
JSC—Johnson Space Center, Houston, Tex.
MSFC—Marshall Space Flight Center, Huntsville, Ala.

CHAPTER IX

Record of Manned Space Flights (Unofficial)

Spacecraft	Country	Astronauts	Launch date	Landing date	Flight time	Remarks
Vostok 1	USSR	Yuri A. Gagarin	April 12, 1961	April 12, 1961	1 hr. 48 min	First manned flight, 1 orbit around Earth.
Mercury-Redstone 3	USA	Alan B. Shepard, Jr.	May 5, 1961	May 5, 1961	0 hr. 15 min	First American in space, suborbital flight.
Mercury-Redstone 4	USA	Virgil I. Grissom	July 21, 1961	July 21, 1961	0 hr. 16 min	Suborbital flight.
Vostok 2	USSR	Gherman S. Titov	Aug. 6, 1961	Aug. 7, 1961	25 hrs. 14 min	More than a day in space.
Mercury-Atlas 6	USA	John H. Glenn, Jr.	Feb. 20, 1962	Feb. 20, 1962	4 hrs. 55 min	First American in orbit, 3 orbits.
Mercury-Atlas 7	USA	M. Scott Carpenter	May 24, 1962	May 24, 1962	4 hrs. 56 min	Three orbits.
Vostok 3	USSR	Andrian G. Nikolayev	Aug. 11, 1962	Aug. 15, 1962	94 hrs. 22 min	Vostok 3 and 4, first group flight—Vostok 4 came within 5 km (3.1 miles) of Vostok 3.
Vostok 4	USSR	Pavel R. Popovich	Aug. 12, 1962	Aug. 15, 1962	70 hrs. 57 min	
Mercury-Atlas 8	USA	Walter M. Schirra, Jr.	Oct. 3, 1962	Oct. 3, 1962	9 hrs. 13 min	Landed 8 km (5 miles) from target.
Mercury-Atlas 9	USA	L. Gordon Cooper, Jr.	May 15, 1963	May 16, 1963	34 hrs. 20 min	First long flight by American.
Vostok 5	USSR	Valery F. Bykovsky	June 14, 1963	June 19, 1963	119 hrs. 6 min	Vostok 5 and 6, second group flight—Vostok 6 came within 4.8 km (3 miles) of Vostok 5.
Vostok 6	USSR	Valentina V. Tereshkova	June 16, 1963	June 19, 1963	70 hrs. 50 min	First woman in space.
Voskhod 1	USSR	Vladimir M. Komarov, Konstantin P. Feoktestov, Dr. Boris G. Yegorov.	Oct. 12, 1964	Oct. 13, 1964	24 hrs. 17 min	First three-man crew.
Voskhod 2	USSR	Aleksei A. Leonov, Pavel I. Belyayev.	Mar. 18, 1965	Mar. 19, 1965	26 hrs. 2 min	First man to "spacewalk" outside spacecraft (Leonov).
Gemini 3	USA	Virgil I. Grissom, John W. Young.	Mar. 23, 1965	Mar. 23, 1965	4 hrs. 53 min	First manned orbital maneuvers.
Gemini 4	USA	James A. McDivitt, Edward H. White, 2nd.	June 3, 1965	June 7, 1965	97 hrs. 48 min	First U.S. "spacewalk," 21 minutes (White).
Gemini 5	USA	L. Gordon Cooper, Jr., Charles Conrad, Jr.	Aug. 21, 1965	Aug. 29, 1965	190 hrs. 56 min	First extended manned flight.

Spacecraft	Country	Astronauts	Launch date	Landing date	Flight time	Remarks
Gemini 7	USA	Frank Borman, James A. Lovell, Jr.	Dec. 4, 1965	Dec. 18, 1965	330 hrs. 35 min.	Longest space flight prior to Soyuz 9.
Gemini 6	USA	Walter M. Schirra, Jr., Thomas P. Stafford.	Dec. 15, 1965	Dec. 16, 1965	25 hrs. 52 min.	Rendezvous within 0.3 meters (1 foot) of Gemini 7.
Gemini 8	USA	Neil A. Armstrong, David R. Scott.	Mar. 16, 1966	Mar. 16, 1966	10 hrs. 42 min.	First docking to "Agena" target, mission cut short.
Gemini 9	USA	Thomas P. Stafford, Eugene A. Cernan.	June 3, 1966	June 6, 1966	72 hrs. 21 min.	Rendezvous, extravehicular activity tests; precision landing.
Gemini 10	USA	John W. Young, Michael Collins.	July 18, 1966	July 21, 1966	70 hrs. 47 min.	Rendezvous with two targets.
Gemini 11	USA	Charles Conrad, Jr., Richard F. Gordon, Jr.	Sept. 12, 1966	Sept. 15, 1966	71 hrs. 14 min.	Rendezvous and docking with Atlas-Agena target vehicle.
Gemini 12	USA	James A. Lovell, Jr., Edwin A. Aldrin, Jr.	Nov. 11, 1966	Nov. 15, 1966	94 hrs. 33 min.	Three successful extravehicular trips.
Soyuz 1	USSR	Vladimir M. Komarov	Apr. 23, 1967	Apr. 24, 1967	26 hrs. 40 min.	Crashed on reentry, killing Komarov.
Apollo 7	USA	Walter M. Schirra, Jr., Donn F. Eisele, R. Walter Cunningham.	Oct. 11, 1968	Oct. 22, 1968	260 hrs. 9 min.	First manned flight of Apollo.
Soyuz 3	USSR	Georgi T. Beregovoy	Oct. 26, 1968	Oct. 28, 1968	94 hrs. 51 min.	Rendezvous with unmanned Soyuz 2.
Apollo 8	USA	Frank Borman, James A. Lovell, Jr., William A. Anders.	Dec. 21, 1968	Dec. 27, 1968	147 hrs.	First manned flight around the Moon.
Soyuz 4	USSR	Vladimir A. Shatalov	Jan. 14, 1969	Jan. 17, 1969	71 hrs. 14 min.	Rendezvous with Soyuz 5.
Soyuz 5	USSR	Boris V. Volynov, Aleksey S. Yeliseyev, Yevgeni V. Khrunov.	Jan. 15, 1969	Jan. 18, 1969	72 hrs. 45 min.	Rendezvous with Soyuz 4; Yeliseyev and Khrunov transfer to Soyuz 4.
Apollo 9	USA	James A. McDivitt, David R. Scott, Russell L. Schweikart.	Mar. 3, 1969	Mar. 13, 1969	241 hrs. 1 min.	Docking with lunar module.
Apollo 10	USA	Thomas R. Stafford, Eugene A. Cernan, John W. Young.	May 18, 1969	May 26, 1969	192 hrs. 3 min.	Descent to within 14.5 km (9 miles) of lunar surface.
Apollo 11	USA	Neil A. Armstrong, Edwin E. Aldrin, Jr., Michael Collins.	July 16, 1969	July 24, 1969	195 hrs. 18 min.	First lunar landing; Armstrong, Aldrin land on Moon.
Soyuz 6	USSR	Georgiy Shonin, Valeriy Kubasov.	Oct. 11, 1969	Oct. 16, 1969	118 hrs. 44 min.	Began first triple launch of manned spacecraft, followed by Soyuz 7 and 8.
Soyuz 7	USSR	Anatoly Filipchenko, Viktor Gorbatko, Vladisov Volko.	Oct. 12, 1969	Oct. 17, 1969	118 hrs. 41 min.	
Soyuz 8	USSR	Vladimir Shatalov, Aleksey Yeliseyev.	Oct. 13, 1969	Oct. 18, 1969	118 hrs. 50 min.	

Mission	Country	Crew	Launch date	Recovery date	Duration	Remarks
Apollo 12	USA	Charles Conrad, Jr., Alan L. Bean, Richard F. Gordon, Jr.	Nov. 14, 1969	Nov. 24, 1969	224 hrs. 36 min	Two EVA's on Moon by Conrad, Bean.
Apollo 13	USA	James A. Lovell, Jr., Fred W. Haise, Jr., John L. Swigert.	Apr. 11, 1970	Apr. 17, 1970	142 hrs. 54 min	Failure of spacecraft oxygen tank about 56 hours into mission caused mission cancellation and return to Earth.
Soyuz 9	USSR	Andrian G. Nikolayev, Vitali Sevastyanov.	June 2, 1970	June 19, 1970	424 hrs. 59 min	Longest space flight to date.
Apollo 14	USA	Edgar A. Mitchell, Stuart A. Roosa, Alan B. Shepard.	Jan. 31, 1971	Feb. 9, 1971	216 hrs. 2 min	Lunar landing, first wheeled lunar vehicle—Modularized Equipment Transporter (hand-pulled cart).
Soyuz 10	USSR	Vladimar A. Shatalow, Alexei S. Yeliseyev, Nikolai N. Rukavishnikov.	Apr. 22, 1971	Apr. 24, 1971	47 hrs. 46 min	Docked with Salyut 1, a prototype manned space station launched unmanned.
Soyuz 11	USSR	Georgiy T. Dobrovolskiy, Valadislov N. Volkov, Viktor I. Patsayev.	June 6, 1971	June 30, 1971	569 hrs. 47 min	Cosmonauts died during reentry.
Apollo 15	USA	James Irwin, Alfred Worden, David R. Scott.	July 26, 1971	Aug. 7, 1971	294 hrs. 44 min	Lunar landing, first use of Lunar Rover Vehicle, a motor-driven astronaut conveyance.
Apollo 16	USA	John W. Young, Thomas K. Mattingly II, Charles M. Duke.	Apr. 16, 1972	Apr. 27, 1972	265 hrs. 51 min	Lunar landing, second mission with LRV.
Apollo 17	USA	Eugene Cernan, Harrison H. Schmitt, Ronald E. Evans.	Dec. 7, 1972	Dec. 19, 1972	301 hrs. 52 min	Lunar landing, third mission with LRV, final Apollo mission, first scientist-astronaut (Schmitt).

CHAPTER X

Bibliography

1. *Skylab Experiment Overview*, Bellcom, Inc., Washington, D.C., January 1, 1971
2. *Skylab In-Flight Experiments, Summary Descriptions*, Skylab Program Office, NASA-George C. Marshall Space Flight Center, Huntsville, Alabama, June, 1971
3. *Skylab—A Manned Scientific Space Laboratory*, a paper presented at the XXIInd International Astronautical Congress, Brussels, Belgium, by Leland F. Belew, NASA-George C. Marshall Space Flight Center, Huntsville, Alabama, September 23, 1971
4. *Skylab—Program Description*, National Aeronautics and Space Administration, Washington, D.C., October, 1971 (available from U.S. Government Printing Office, Washington, D.C., Price 60 cents, Stock Number 3300–0411)
5. *Skylab—Experiment Integration Summary*, NASA-George C. Marshall Space Flight Center, Huntsville, Alabama, 1972.
6. *Skylab Experiments*, National Aeronautics and Space Administration, Office of Manned Space Flight, Washington, D.C., August, 1972.

In addition to these documents, a large number of reports, notes, memoranda, and press kits were used which provided up-to-date information on Skylab.

CHAPTER XI

Glossary

ABLATION—The removal of surface material from a body by vaporization, melting, chipping, or other erosive process; specifically, the intentional removal of surface matter from a reentry body during high-speed movement through a planetary atmosphere to protect the remaining body from the heat generated by friction.

ABSORPTIVITY—A property of a material, characterizing its capability to absorb rather than transmit or reflect incident radiant energy.

ALBEDO—Relative brightness of a surface or region, measured as the ratio of the amount of reflected radiation to the amount of incident radiation.

ALDOSTERONE—The principal electrolyte-regulating steroid secreted by the adrenal complex (steroid is a group name for compounds that resemble cholesterol chemically). Some of the substances in this group include sex hormones and bile acids.

ALPHA PARTICLE—The positively charged nucleus of a helium atom. Each alpha particle consists of two protons and two neutrons. Alpha particles are emitted by some of the radioactive substances; they are also found in cosmic radiation.

ALTIMETER—An instrument that determines height above a reference level, commonly by measuring the change of atmospheric pressure, or by measuring vertical distance directly with a radar-type system.

ANGIOTENSIN—A vessel-constricting substance present in the blood, and formed by the action of renin (enzyme involved in changing proteins into other products) on a globulin (a class of proteins characterized by being insoluble in water, but soluble in saline solutions).

ÅNGSTROM UNIT—Unit of length employed to measure wave lengths of light. One Ångstrom unit is one ten-thousandth of a micrometer or one ten-billionth of a meter. Ten Ångstrom units make one nanometer.

ANTIBODY—A substance produced by the body in response to the introduction of a foreign substance (antigen).

ANTIDIURETIC HORMONE (vasopressin)—A hormone that suppresses the secretion of urine. Vasopressin also stimulates the contraction of the muscular tissue of the capillaries and arterioles.

ANTIGEN—A substance which, when introduced into the body, stimulates the production of antibodies. Bacteria, their toxins, red blood corpuscles, tissue extracts, pollens, dust, and many other substances may act as antigens.

APOGEE—That point on the trajectory of an Earth-orbiting body which is most distant from the Earth. Also used in connection with orbits of celestial bodies around other celestial bodies.

BALMER SERIES—The visible line spectrum emitted by hydrogen. The wavelengths of the lines form a series, the formula of which was established by Balmer (1885).

BANDPASS FILTER—A wave filter that has a single transmission band extending from a lower cutoff frequency greater than zero to a finite upper cutoff frequency.

BIOASSAY—Estimation of the active power of a sample of a drug, or of other agents or influences, by noting their effects on animals or man.

BLOOD PLASMA—The liquid component of the blood in which the corpuscles are suspended. Plasma may be obtained from whole blood by removing the corpuscles by centrifuging or by sedimentation. It contains all the chemical constituents of whole blood except hemoglobin.

BORESIGHTING—A process of parallel alignment of the lines of sight of two instruments by an optical procedure.

CARDIOTACHOMETER—An instrument for counting or recording the heart beats over long periods of time.

CARDIOVASCULAR—Heart and vessel system.

CHROMOSPHERE—Layer of the solar atmosphere, about 14,000 km (8000 naut. mi.) thick, which surrounds the Sun's visible surface (photosphere). It is best observable during an eclipse or other occultation of the solar disc.

CIRCADIAN RHYTHM—A rhythm with a period of about 24 hours, applied especially to the rhythmic repetition of certain phenomena in living organisms at about the same time each day.

CONTROL MOMENT GYROSCOPE—A large and heavy gyroscope suspended by a two-axis gimbal system. The outer gimbal axis is connected with the spacecraft through a torque motor, the inner axis is free to precess. By energizing the torque motor, the spacecraft attitude can be controlled.

CONVECTION—Mass motions with a fluid (liquid or gas) resulting in transport and mixing of the components of that fluid. Thermal convection results from temperature differences within the fluid.

CORONA—The tenuous envelope of the Sun, beginning about 14,000k m (8000 naut. mi.) above the solar surface and extending many millions of kilometers into space. The corona is visible only when the solar disc is occulted.

CRAWLER—A large tracked vehicle also called Transporter. It is similar to machines used in strip mining operations. The vehicle moves on four double tracks and transports the Saturn V with the Skylab and mobile launcher, and also the Saturn 1B, from the Vertical Assembly Building to the launch pad at the Kennedy Space Center.

CYTOGENETICS—The branch of genetics devoted to the study of the cellular constituents which are concerned with heredity (chromosomes and genes). Also, the scientific study of the relationship between chromosomal aberrations and pathological conditions.

DECIBEL—The decibel is a ratio of two numbers, equal to the tenth root of ten or about 1.259. It is mostly used as the ratio of two power levels. If a radio receiver, playing at a certain "volume" level, is turned up until the acoustic power output has increased by 25.9%, its new level is one decibel above the original level.

DEGREES K—Degrees Kelvin. A degree of temperature on the Kelvin scale, also called "absolute scale." The Kelvin zero point is approximately $-273.1°$ Centigrade. A degree Kelvin is equal in magnitude to a degree on the Centigrade scale.

DEHYDROGENASE—An enzyme which mobilizes the hydrogen of a substrate (the base on which an organism lives) so that it can pass to a hydrogen acceptor.

DESSICANT—A drying agent.

DICHROIC—The property of a substance to appear in one color by reflected light and in another by transmitted light.

DIFFUSION—In an atmosphere, as in any gaseous system, the exchange of fluid parcels between regions, in apparently random motions of a scale too small to be treated by the equations of motion.

DNA—Deoxyribonucleic acid, a complex organic acid of high molecular weight consisting of chains of alternate units of phosphate and a pentose sugar (a sugar having five oxygen atoms) which has a purine and pyrimidine base attached to it. In DNA the sugar is 2-deoxyribose. DNA is believed to carry all the hereditary traits of a species coded in the sequence of atomic groups along its length. See RNA.

DOSIMETER—An instrument for measuring the accumulated flux of particle or photon radiations, such as protons in the Van Allen Belts, or X-rays in solar radiation.

ELECTROCARDIOGRAM—A written or printed record of the heart's action, made by an electrocardiograph, an instrument for recording the changes of electrical potential occurring during the heartbeat.

ELECTROLYTE—A solution which conducts electricity. Passage of current is accompanied by liberation or consumption of matter at the electrodes. Also, a substance, as an acid, base, or salt, that becomes such a conductor when dissolved in a suitable solvent, or fused. The current is carried by charged particles (ions).

ELECTROPHORESIS—The movement of molecules or other very small particles through a fluid under the action of an external electric field. Positively charged particles (metallic oxides, basic dyestuffs) migrate to the cathode, and negatively charged particles (metals, sulfur, metallic sulfides, acid dyestuffs) migrate to the anode.

EMISSIVITY—A property of a material, characterizing its capability to emit electromagnetic radiation as a consequence of its inherent thermal energy.

ENDOCRINE GLAND—A ductless gland whose secretions pass directly into the lymph or blood stream. These glands produce hormones which control action and development of other parts of the body either by activation or by inhibition.

ENZYME—An organic compound, frequently a protein, capable of accelerating or producing by catalytic action some change in an organic substance for which it is often specific. An activating enzyme activates a given amino acid by attaching it to the corresponding transfer ribonucleic acid.

EPHEMERIS—Periodical publication which lists the predicted positions of celestial bodies at regular intervals, and the times of astronomical occurrences. It also contains other data of interest to astronomers.

EPINEPHRINE CONCENTRATIONS—Concentrations of adrenaline (compound occuring naturally as the adrenal hormone). In certain concentrations the compound causes an increase in blood pressure and in the sugar content of the blood.

ERGOMETER—A device for measuring energy expended or work done.

EUTECTIC—In certain ranges of metal alloys there is one mixture which melts at a lower temperature than any other alloy in the series. Such an alloy is termed the eutectic.

EXOTHERMIC—A process which releases, rather than absorbs, heat energy.

FABRY-PÉROT INTERFEROMETER—An instrument utilizing the wave nature of light to cause constructive or destructive interference of light by passing a light beam between two parallel, partially reflecting surfaces.

FIBRIN—A whitish, insoluble protein which forms the essential portion of the blood clot.

FIBRINOLYSIS—The splitting up of fibrin by enzyme action.

FIBRINOLYTIC—Pertaining to, characterized by, or causing fibrinolysis.

FILAMENT—In solar physics, filament designates a quiet prominence or plasma cloud high´above the chromosphere, visible as a dark narrow patch against the bright solar disc.

FLARE—A violent eruption on the Sun's surface (chromosphere), accompanied by emissions of protons and other particles and of electromagnetic radiation.

FRAUNHOFER LINES—Dark lines in the spectrum of solar radiation, produced by the absorption of light by gases in the outer portions of the Sun.

GEGENSCHEIN—A faint light observed from the dark side of the Earth in a direction opposite to the Sun. It results from sunlight reflectd by dust particles which orbit the Sun at planetary distances.

GEOCORONA—That region around the Earth which extends from a height of about 600 km (330 naut. mi.) to about 3000 km (1700 naut. mi.). It consists mainly of helium in the lower regions, and of hydrogen in the upper regions.

GIMBAL—A ring or frame with two mutually perpendicular and intersecting axes of rotation, providing free angular movement in two directions, on which a rocket engine, or a gyroscope, or another instrument may be mounted.

GLUCOSE—Chemical name for a natural sugar found in fruits and in the blood.

GLUTATHIONE—Co-enzyme of glyoxalase which acts as a respiratory carrier of oxygen.

GLYCERALDEHYDE—A compound formed by the oxidation of glycerol, a mixture of glycerin and acetanilid powder.

GRANULATION—A net-like pattern of irregular cells on the solar surface, visible in white light. The cells are caused by plasma convection within the photosphere. Cell diameters are 800 km (450 naut. mi.) to 3000 km (1700 naut. mi.); each individual cell has a lifetime of several minutes.

GREENWICH MEAN TIME—Mean solar time at the Greenwich meridan, used by most navigators, and adopted as the prime basis of standard time throughout the world.

GROUND TRUTH—Definition of Earth surface conditions through direct measurements or visual inspections for calibration or evaluation of remote sensing observations made from satellites or aircraft.

HEMATOCRIT—An instrument for determining the relative amounts of plasma and corpuscles in blood, generally some form of centrifugal apparatus.

HEMATOLOGY—A branch of medicine concerning the study of the blood, the blood-forming tissues, and the diseases of the blood.

HEMOLYSIS—The dissolution of red blood corpuscles with liberation of their hemoglobin, an iron-containing protein respiratory pigment occurring in the red blood cells.

HEXOKINASE—An enzyme that catalyzes the transfer of a high-energy phosphate group of a donor to D-glucose, producing D-glucose-6-phosphate.

HUMORAL—Pertaining to fluid or seimfluid substances in the body.

IMMUNOLOGY—The medical, bacteriological, and chemical study of the phenomena and causes of immunity.

IONOSPHERE—Region of ionized gases surrounding the Earth and extending from about 60 km (33 naut. mi) to distances up to several hundred kilometers. The ionization in this region is due to bombardment by ultraviolet radiation and X-rays from the Sun, and by cosmic rays. The existence of this region makes possible long-range radio communications through reflection of terrestrial radio transmission.

ISOTOPE—Isotopes of an element have the same number of protons in their nuclei, and hence represent the same element, but they differ in the number of neutrons and therefore in mass number.

KINASE—An organic substance (enzyme) which activates other substances to develop into chemical ferments or enzymes.

LEUKOCYTE—White blood corpuscle.

LIMB—The outer edge or a portion of the edge of a celestial body such as the Moon, Venus, or the Sun, as seen from a distance.

LYIDE (Lipid)—A fatty acid insoluble in water.

LYMAN CONTINUUM—The ultraviolet region of the spectrum of the hydrogen atom immediately adjacent to the Lyman line spectrum with wavelengths shorter than 912 Å.

LYMPHOCYTE—Lymph cell or white blood corpuscle without cytoplasmic granules.

MAGNETOSPHERE—The region around the Earth above about 160 km (90 naut. mi.) and below the magnetopause (about 15 earth radii in the solar direction, and at least 40 earth radii in the antisolar direction). Inside the magnetosphere, the Earth's magnetic field is dominant; outside, the interplanetary magnetic field dominates.

MASS SPECTROMETER—An instrument which determines the masses of atoms and molecules.

METABOLIC CHANGES—The sum of all physical and chemical changes which take place within an organism; all energy and material transformations which occur within living cells.

METABOLISM—The interchange of materials between living organisms and the environment, or within a living organism, by which energy for maintaining life is secured.

METAPHASE—The stage during cell division in which the chromosomes are arranged in an equatorial plate.

METEOR—The light resulting from the transition of a solid particle (meteoroid) from space through the Earth's atmosphere, commonly called a "shooting star" or "falling star."

METEORITE—A solid particle from space which enters the Earth's atmosphere and reaches the surface. Meteorites are classified as iron meteorites (siderites) and stone meteorites (aerolites) according to their compositions.

METEOROID—A solid object moving through interplanetary space of a size considerably smaller than an asteroid and considerably larger than an atom or molecule. When the object glows while traveling through the Earth's atmosphere, it is called a meteor; when it reaches the surface of the Earth, it is called a meteorite.

METHEMOGLOBIN—A soluble brownish-red, crystalline compound from which the oxygen cannot be removed in a vacuum. It is formed by the spontaneous decompasition of blood, and also by the action on blood of various oxidizing reagents, as oxone, etc.

MICRON—A unit of length equal to one-millionth of a meter or one-thousandth of a millimeter, usually called micrometer.

MORPHOLOGY—Branch of biology dealing with the form and structure of animals and plants. It includes anatomy, histology and organography, and also the non-physiological aspects of cytology and embryology.

MULTISPECTRAL—Utilizing radiation from several discrete bands of the spectrum simultaneously.

NADIR—The direction vertically downward (opposite to zenith).

NANOMETER—One-billionth of a meter (10^{-9} meter).

NOREPINEPHRINE—A hormone secreted by the adrenal medulla in response to stimulation in the viscera, and stored in granules that stain strongly with chromium salts. Granules are released predominantly in response to hypotension (diminished tension or lower blood pressure).

OCCULTATION—The disappearance of a celestial body behind another body of equal or larger apparent size, such as the occultation of the Sun by the Moon as viewed by an Earth observer during a solar eclipse. Also, the covering of the image of a celestial body by a disc whose size is equal to or larger than the size of the image.

OCULOGYRAL ILLUSION—An illusion developed by the movement of the eye about an axis from the front to the rear of the head.

ORTHOSTATIC—The upright, or erect, position of the human body.

OSMOLALITY—The property of a liquid to exercise an osmotic pressure because it contains an electrolyte in solution.

OUTGASSING—The emanation of gas from a material. This process is usually enhanced in vacuum.

PERIGEE—That point on the trajectory of an orbiting body which is nearest the Earth when the Earth is the center of attraction. Also, used in connection with orbits of celestial bodies around other celestial bodies.

PHOTOMETER—An instrument for measuring the intensity of light by comparing it with a standard.

PHOTOMULTIPLIER TUBE—An electron tube that produces electrical signals in response to light. In the tube, the electric signal is amplified to produce a measurable output signal even from very small quantities of light.

PHOTOSPHERE—Intensely luminous surface layer of the Sun in which the sunspots and several other solar phenomena occur.

PLAGES—Bright regions in the chromosphere of the Sun, usually near sunspots, indicating areas of enhanced magnetic field strength (10 to 100 Gauss*) and increased solar activity. Plages are best observed in monochromatic light of hydrogen or calcium.

PLASMA—A gas composed of ions, electrons, neutral atoms, and molecules. The interactions between particles of a plasma are mainly electromagnetic. Although many of the individual particles are electrically positive or negative, the plasma as a whole is neutral.

PLASMA RENIN—Plasma is the fluid portion of the blood in which the corpuscles are suspended. Renin is an enzyme involved in changing proteins into other products.

PLETHYSMOGRAPH—An instrument for determining and registering variations in the size of an organ part or limb, and in the amount of blood present or passing through it.

POLARIMETER—An instrument for determining the degree of polarization of electromagnetic radiation, specifically the polarization of light.

POLARIZATION—The state of electromagnetic radiation when the transverse oscillations take place in some regular manner, e.g., all in one plane.

PRECESSION—The angular motion (tilting) of the axis of a spinning or rotating body caused by a torque whose axis is not parallel with the axis of rotation.

PRINCIPAL INVESTIGATOR—The researcher responsible for defining an investigation or experiment as part of a NASA science or technology project, and for ensuring that the investigation of the experiment meets with its specific objectives.

PROMINENCE—A plasma protuberance above the surface of the Sun, originating in the chromosphere and extending sometimes to a height of several hundred thousand kilometers.

---

*For comparison, the Earth's magnetic field near the surface is 0.6 Gauss.

Proton—Positively-charged nuclear particle which forms a significant part of all atomic nuclei. The nucleus of a normal hydrogen atom is a proton. It is 1,837 times heavier than the electron.

Radiometer—An instrument for detecting and measuring radiant energy.

Raster—A geometric pattern followed by the sending element of a detector system, or by the electron beam of a television transmitter or receiver.

Rate Gyro—A gyroscope with one free gimbal axis (precession axis). Motion around this axis is constrained by a spring. The second gimbal axis is tied to the spacecraft. An angular motion of the spacecraft around this axis produces an angular excursion of the precession axis against the spring force; the angle of excursion is proportional to the rate of angular rotation in radians per second.

Reductase—An enzyme that has a reducing action on chemical compounds.

Redundancy—Originally "exceeding what is necessary or normal;" technically a back-up system which takes over when the prime system fails.

Resolving power (light)—The ability of an optical system to separate adjacent points and lines in the image and to show fine detail of the target.

RNA—Ribonucleic acid, a nucleoprotein found in the cell's cytoplasm. It probably controls protein synthesis under the regulatory influence of DNA. See DNA.

S-band—A range of frequencies used in radar and communications that extends from 1,550 to 5,200 megahertz (wavelength region from 0.06 m to 0.2 m).

Scintillation—A flash of light produced in a phosphor by an ionizing particle or photon.

Single crystal—A crystal having a homogeneous, undisturbed lattice structure throughout.

Slew—To change the direction of an antenna or telescope in order to follow a moving target, or to change from one target to another.

Solar wind—Streams of particles (mostly ions of hydrogen and helium, and electrons) emanating from the Sun and flowing approximately radially outward at velocities from 300 to 800 km per sec.

Spectroscope—An optical instrument which spreads a beam of electromagnetic radiation into a spectrum of different wavelengths for visual inspection.

Spectrograph—Modification of a spectroscope in which the spectrum is photographed or recorded electronically.

Spectroheliograph—A modification of the spectrograph which permits taking pictures of the complete solar disk in monochromatic light.

Spirometer—An instrument to measure the breathing volume.

Sunspots—Dark, irregular areas on the solar surface with strong magnetic fields (1000 to 2000 Gauss) surrounded by gray zones (penumbra). Sunspots occur mostly in pairs, with opposite magnetic polarity; they persist for periods of days, weeks, or months. Their temperatures are about 2000° K lower than the temperature of the surrounding photosphere (5780° K). The frequency of sunspots follows an eleven-year cycle.

Supergranulation—A net-like pattern of irregular cells, 15,000 km (800 naut. mi.) to 40,000 km (22,000 naut. mi.) in diameter, superimposed upon the photospheric granulation network, and caused by plasma convection within the chromosphere. Cell lifetime is about 20 hours. Supergranulation is best observed in the monochromatic light of a calcium spectral line.

Synoptic—Observing different objects, or different aspects of one object, at the same time.

Syringe—An instrument for injecting or extracting liquids into or from any vessel or cavity, such as a vein in the human body.

Terrestrial—Of or pertaining to the Earth.

Toxin—A soluble poison produced and liberated by certain bacteria, insects, snakes, and plants. Toxins are usually protein substances which may be destroyed by heat.

Vectorcardiograph—An instrument for taking a graphic record of the magnitude and direction of the electrical potentials of the heart.

VELCRO—A fastener for quick attachment (and subsequent detachment) of an object to another object, or to a surface. It has two parts, a pad consisting of velvety, loopy pile made of Teflon, and a pad consisting of little hooks resembling a cockleburr, made of polyester. Attachment by pressing the two pads together, detachment by pulling them apart.

VESTIBULAR—Pertaining to the organs of the inner ear that provide a sense of equilibrium for animals and man.

VIDEO—Pertaining to the picture signals in a television system, or to the information-carrying signals which are eventually presented on the cathode ray tubes of a radar system.

ZEOLITE—A group of hydrous silicates of aluminum, containing sodium and calcium, and giving off water upon heating; technically, a substance capable of absorbing large quantities of carbon dioxide.

ZODIACAL LIGHT—A faint light emanating from a region in the night sky roughly defined by the zodiac. This light results from sunlight scattered by fine dust particles that orbit the Sun at planetary distances; it is enhanced in the vicinity of the Sun.

CHAPTER XII

Acronyms

AM—Airlock Module; also Amplitude Modulation.
APCS—Attitude and Pointing Control System.
ASMU—Automatically Stabilized Maneuvering Unit.
ATM—Apollo Telescope Mount.
ATMDC—Apollo Telescope Mount Digital Computer.
BMMD—Body Mass Measurement Device.
CM—Command Module.
CMC—Command Module Computer.
CMG—Control Moment Gyro.
COAS—Crewman Optical Alignment Sight.
CS—Crew Station.
CSM—Command and Service Module.
DA—Deployment Assembly.
DAC—Data Acquisition Camera.
db—Decibel.
DCS—Digital Command System.
DNA—Deoxyribonucleic Acid.
DSKY—Display Keyboard.
DT—Delayed Time.
ECS—Environmental Control System.
EEG—Electroencephalogram.
EPS—Electrical Power System.
EREP—Earth Resources Experiment Package.
ESS—Experiment Support System.
EVA—Extravehicular Activity.
FAS—Fixed Airlock Shroud.
FM—Frequency Modulation.
GMT—Greenwich Mean Time.
GSFC—Goddard Space Flight Center.
HCO—Harvard College Observatory.
HHMU—Hand Held Maneuvering Unit.
HSS—Habitability Support System.
IU—Instrument Unit.
IVA—Intravehicular Activity.
JSC—Johnson Space Center.
KSC—Kennedy Space Center.
LO—Liftoff.
LSU—Life Support Umbilical.
LV—Launch Vehicle.
MCC—Mission Control Center.
MDA—Multiple Docking Adapter.
MSFC—Marshall Space Flight Center.
NASA—National Aeronautics and Space Administration.
NRL—Naval Research Laboratory.
OMSF—Office of Manned Space Flight.
OWS—Orbital Workshop.
PCM—Pulse Code Modulation.

PI—Principal Investigator.
PS—Payload Shroud.
RNA—Ribonucleic Acid.
RPM—Revolutions Per Minute, also Roll Positioning Mechanism.
SAL—Scientific Airlock.
SAS—Solar Array System.
S/C—Spacecraft.
SL—Skylab.
SM—Service Module.
STDN—Spacecraft Tracking and Data Network.
TACS—Thruster Attitude Control System.
TCS—Thermal Control System.
TM—Telemetry.
TV—Televison.
UV—Ultraviolet.
VHF—Very High Frequency.

U.S. GOVERNMENT PRINTING OFFICE : 1973—O-500-721

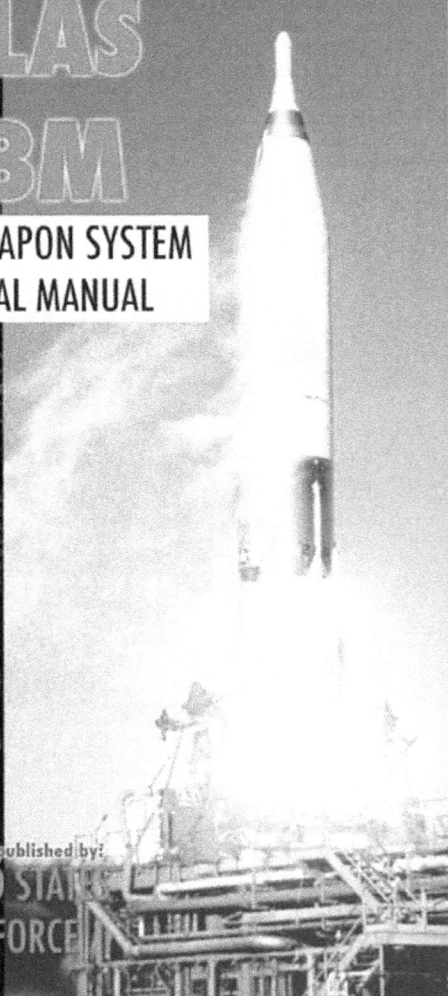

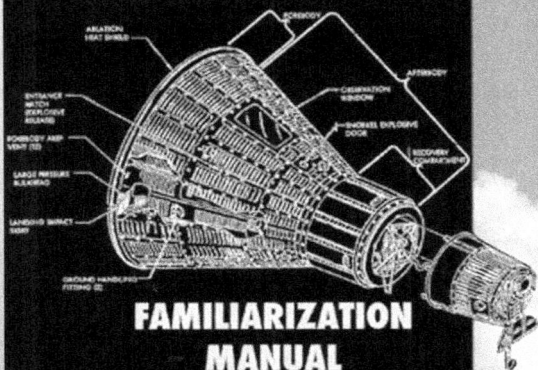

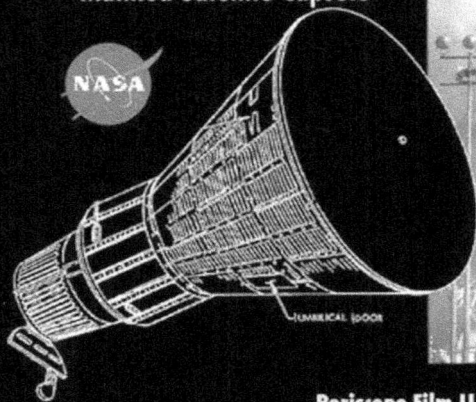

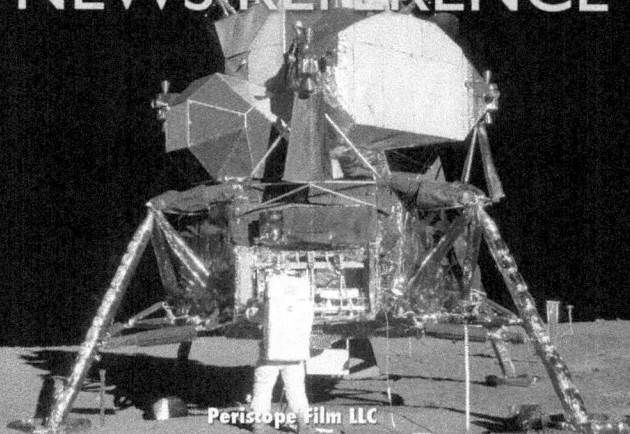